T0178251

From Plant Traits to Vegetation Structure

Plant community ecology has traditionally taken a taxonomical approach based on population dynamics. This book contrasts such an approach with a trait-based approach. After reviewing these two approaches, it then explains how models based on the Maximum Entropy Formalism can be used to predict the relative abundance of different species from a potential species pool. Following this it shows how the trait constraints, upon which the model is based, are necessary consequences of natural selection and population dynamics. The final sections of the book extend the discussion to macroecological patterns of species abundance and conclude with some outstanding unresolved questions. Written for advanced undergraduates, graduates and researchers in plant ecology, Bill Shipley demonstrates how a trait-based approach can explain how the principle of natural selection and quantitative genetics can be combined with maximum entropy methods to explain and predict the structure of plant communities.

BILL SHIPLEY obtained his PhD in plant ecology from the University of Ottawa in 1987 and now teaches plant ecology and statistics at the Université de Sherbrooke (Qc) Canada. He is author of over 70 peer-reviewed papers in ecology and statistics and *Cause and Correlation in Biology: A user's guide to path analysis, structural equations, and causal inference* (Cambridge University Press, 2000).

From Plant Traits to Vegetation Structure

Chance and selection in the assembly of ecological communities

BILL SHIPLEY

Université de Sherbrooke, Sherbrooke (Qc) Canada

CAMBRIDGE UNIVERSITY PRESS

CAMBRIDGE
UNIVERSITY PRESS

Shaftesbury Road, Cambridge CB2 8EA, United Kingdom

One Liberty Plaza, 20th Floor, New York, NY 10006, USA

477 Williamstown Road, Port Melbourne, VIC 3207, Australia

314–321, 3rd Floor, Plot 3, Splendor Forum, Jasola District Centre, New Delhi – 110025, India

103 Penang Road, #05–06/07, Visioncrest Commercial, Singapore 238467

Cambridge University Press is part of Cambridge University Press & Assessment, a department of the University of Cambridge.

We share the University's mission to contribute to society through the pursuit of education, learning and research at the highest international levels of excellence.

www.cambridge.org
Information on this title: www.cambridge.org/9780521133555

First published 2010

A catalogue record for this publication is available from the British Library

ISBN   978-0-521-11747-0   Hardback
ISBN   978-0-521-13355-5   Paperback

Cambridge University Press & Assessment has no responsibility for the persistence or accuracy of URLs for external or third-party internet websites referred to in this publication and does not guarantee that any content on such websites is, or will remain, accurate or appropriate.

À Lyne, qui garde mes pieds sur terre
quand ma tête est dans les nuages.

Contents

Contents

Preface

*"If you can look into the seeds of time, and say which grain will grow
and which will not, speak then to me."*

Banquo, from Shakespeare's Macbeth

Perhaps this book is my attempt to hold a conversation with Banquo? He
would be a fierce critic of the scientific stature of plant ecology, judging
from this quote, and I doubt that I could get him to speak at all. He
demands, not only predictive ability, but a very fine-grained level of
predictive ability. In this book I attempt to develop a predictive theory of
plant community assembly. By "predictive" I mean that the theory should be
able to quantitatively tell us the relative abundance of each species in the
local community under natural field conditions and based on information that
can be collected in practice. By "theory" I mean a formal method of both
performing such predictions and also of being able logically to deduce why
such predictions actually hold in nature based on known biological processes
– in this case, the process of natural selection.

I warn you at the outset that the predictive theory presented in this book
would not satisfy Banquo even if it were to succeed. Such fine-scale
predictive ability – being able to "say which grain will grow and which will
not" – is likely forever beyond our grasp. The reason for this is explained in
Chapter 2. We must remain forever mute before Banquo. However, this does
not mean that macroscopic properties of plant to communities must remain
forever beyond our grasp. We might be able accurately to predict such
aggregate properties as the relative abundance of various species even if we
can never say what will happen with every seed. In fact, counterintuitive as
this might sound, such inherent unpredictability at a fine scale might even
help us to predict such macroscopic properties. This is the message of
Chapters 4 to 7. On reflection then, my conversation cannot be with Banquo
but rather with you, the reader.

So who are you? I find it easier to write if I can imagine the person to whom I am speaking and then simply write down the conversation. Let me tell you who I imagine you to be. You are a graduate student of ecology, or perhaps an advanced undergraduate student, who is interested in how ecological communities are formed. No? Then you are a more seasoned researcher in plant ecology who is looking for a new perspective on a question as old as our science. That's not you either? Then you are an animal ecologist, or maybe an ecologist without any taxonomic preference, who is looking for a *really* new perspective on the same old question. If this is who you are then I dearly hope that you will apply these ideas to your own subject area. Perhaps your interests are more theoretical? If so you will also find some new theoretical developments that might interest you. In case you are *not* a theoretical ecologist I hasten to add that explicitly mathematical developments have been separated from the main text by placing them inside boxes. You can certainly read the text – if you want – without wading through these boxes. You will have to endure some equations but you will not need any mathematics beyond basic algebra, some introductory calculus, and the rudiments of statistics. Although not a user's manual, you should find enough practical information (including some code in the *R* language) to be able to apply the ideas of this book to real field data.

Ideas are only scientific ones when they are made public. Public ideas (memes) are ones that travel from brain to brain. Each passage through some grey matter changes the idea, sometimes profoundly but usually only subtly, and this gradual change makes it very hard to say where an "idea" originates or even if two people are thinking about the same "idea". The ideas in this book are no exception. I am constantly surprised when I find one of my own apparently original ideas written, long before I ever thought it, in the writings of people who have influenced me. Properly chastened, such surprises remind me both of how much I owe to others and also of how few original ideas I actually have. I have tried to properly credit such borrowed ideas but, just to be sure, I want to acknowledge my debt to the following people who have each influenced my thoughts on the subject: Robert van Hulst, Paul Keddy, Rob Peters, Martin Lechowicz, Len Lefkovitch and Éric Garnier. Grégory Sonnier (a current PhD student) has been working hard to find new ways of empirically testing many of the ideas in this book. Some of his interesting results have not been included in this book since they have not yet been published, but they have undoubtedly influenced me as well. If anyone notices that I have not properly acknowledged the ideas of others I can only ask that they let me know so that I can properly pay my debts.

Bernard Colin, Robert van Hulst, Martin Lechowicz, Dany Garant, Roderick Dewar, Bruce Glymour, Daniel Laughlin, Rafael Otfinowski and Grégory Sonnier have read and commented on some of the chapters. Thank you all.

1

Playing with loaded dice

Analogies are a vital part of science because they help us to imagine the unknown with reference to that which we already understand. Although vital, analogies become dangerous when they become so entrenched that we confuse the analogy with reality. The danger, when this happens, arises because we are prevented from conceiving of nature in any other way. In medieval discourse the organizing Aristotelian analogy was *nature-as-an-organism*. Natural phenomena were seen to possess a life cycle: birth, growth, old age and death. Processes in the natural world were made intelligible in this way and were understood by comparison with the inherent desire (the "nature") of sentient organisms to attain goals (teleology). Cats hunt mice. Why? Because a cat is a predator and it is in the nature of predators to hunt. A stone falls to the ground rather than flying up into the air. Why? Because it is in the nature of heavy objects (i.e. objects made of "earth" rather than "air") to move down. To know the nature of a thing was to know the thing itself (Dear 2007).

This analogy was replaced in the seventeen century, by people like Descartes, Galileo and Newton, with a new one: the analogy of *nature-as-a-machine*. The hand of a mechanical clock doesn't move around the face because it is in its "nature" to do so – place the hand alone on a table and it remains stationary. Rather that an inherent desire, the hand, when properly installed, moves because a spring turns a wheel that hits a cog. Sometime during the seventeenth century Nature stopped being teleological and became inert matter in motion obeying physical causes. By the end of the seventeenth century the very meaning of "explanation" itself had changed from its original Aristotelian one to the new mechanistic one: to know the cause–effect sequence was to know the thing.

What is true for great scientific ideas is also true of more modest ones because these too are more easily grasped through analogy. As will be explained in Chapter 2, the analogy used by ecologists to understand

1

community assembly has been the dominant one of nature-as-a-machine. This analogy is useful as long as the cause–effect sequences are relatively few in number and constant over time. However, when dealing with large numbers of organisms in interaction with each other and with the physical environment, the cause–effect sequences are so numerous, complicated and contingent that the nature-as-a-machine analogy becomes a hindrance rather than a help. We must constantly remind ourselves that Nature is neither a machine nor an organic entity. Nature is simply Nature. We must not confuse the analogy with the thing.

The analogy used in this book for the process of community assembly is *nature-as-a-biased-die*. The resulting model is one of community assembly through chance constrained by natural selection. It is, as John Herschel[1] might have called it, a model of community assembly by "higgledy-piggledy".

In this book I compare the process of community assembly to a never-ending game of crooked dice. Imagine a large hall, thick with cigar smoke, and filled with tables of different shapes and made of different materials. There are steel tables, ones make of foam, and ones stained with Maple syrup. There are tables that are slanted and ones that vibrate or move unexpectedly. The tables are the different physical environments upon which community assembly occurs. This assembly occurs as players constantly throw millions of many-faced dice onto these tables, placing bets, and winning or losing resources. Each face represents a different species and every time a die on a given table presents a particular face the corresponding species wins some unit of resource and so increases its abundance in the corresponding environment. Most importantly, the dice are biased by different physical properties of each face (i.e. functional traits of species) such that different species have different probabilities of winning or losing on different tables. For instance, a face might be magnetized. Whether this helps or hinders the associated species in this analogy depends on whether the table (the environment) is made of steel or wood, or if the table itself is magnetized positively or negatively, or even if faces of other dice are magnetized as well.

---

[1] Sir John Frederick William Herschel FRS (1792–1871) was one of his generation's top English scientists. In his *Physical Geography of the Globe* (Herschel 1872) he made the condenscending description of Darwin's notion of natural selection as the "law of higgledy-piggledy" and stated: "We can no more accept the principle of arbitrary and casual variation and natural selection as a sufficient account, *per se*, of the past and present organic world, than we can receive the Laputan method of composing books (pushed à *l'outrance*) as a sufficient one of Shakespeare and the *Principia*." The irony is that the "Laputan" method to which Herschel refers (i.e. writing books by randomly combining words) comes from Johnathan Swift's *Gulliver's Travels* and refers to the one used by the absurd and useless philosophers inhabiting the island of Laputa. This tale was a parody of the Royal Society of whom Herschel was a member.

The dynamics of this game of crooked craps is the dynamics of community assembly. I think that I know what you are thinking: How can we possibly develop predictive models of community assembly from such a strange image? Such a view of ecology might seem bizarre right now but, I hope, it will seem less strange when you reach the end of this book.

This book is not about plant traits even though the notion of a plant trait is central to it. There are many good books dealing with physiological (Lange *et al.* 1983, Pearcy *et al.* 1991, Lambers *et al.* 1998, Larcher 2001) or functional ecology (Grime 1979, Hendry and Grime 1993, Grime 2001, Grime 2007) but this book is not one of them. Chapter 3 outlines some of the key empirical results linking the functional traits of plants to community structure, but it is certainly not an exhaustive review of functional plant ecology. The notion of vegetation structure – of which species are present at a site, at what abundances, and how these two properties of vegetation change with environments – is also central to this book but I did not write a book about vegetation structure. Rather, this book deals with how vegetation structure emerges from the stochastic interplay of the functional traits of individual plants with environmental gradients and how we can quantitatively predict the resulting community structure that is generated by this interplay. You will be introduced to a new mathematical model of community assembly but the empirical knowledge of functional plant ecology will dominate the mathematical formalism; this book is definitely not a treatise on mathematical modeling in ecology and no advanced mathematical ability is needed.

So what is this book? It is *methodological* but only if "methodology" is understood to include its conceptual and philosophical components as well as its "how-to" aspects. It is *theoretical* but only if theory is understood in its most general context of converting biological assumptions into ecological conclusions through logical inference. It is *empirical* but only in the sense that the assumptions of the logical inference are justified by repeatable and quantifiable patterns between plant traits and environmental gradients that are measured in the field. Finally it is *synthetic* since the goal is to integrate each of the approaches mentioned in this paragraph.

My thoughts about the nature of this interplay first began to take form over 20 years ago during my doctoral studies but they only began to form a unified whole during the writing of this book. To my surprise I found the conceptual link between traits, plants and communities in one of the most fundamental notions of biology: natural selection. However, to appreciate this link, and especially to cast it in a more rigorous and mathematical form, I first had to realize that I (and most other biologists) have been wearing blinders with respect to our view of natural selection. Ever since Darwin and Wallace first worked out the process of natural selection we have been trained to mentally associate "natural selection" with "evolution". This is perfectly understandable; the primary interest of both Darwin and Wallace was, not natural selection itself, but rather its evolutionary consequences. For Darwin and Wallace natural selection was a means to a more important end.

Natural selection is a process which, when operating between genotypes belonging to the same interbreeding group, *leads* to evolution by adaptation. Natural selection is one of the causes of evolution. No modern biologist would make the mistake of claiming that natural selection *is* evolution. We all understand that there are other processes besides natural selection that contribute to evolutionary divergence. However, most biologists have made the mistake of assuming that natural selection, though not the *only* cause of evolution, is *only* a cause of evolution; that is, that natural selection has no biological consequences besides those related to evolution. Because of this it seems strange to inquire about the consequence of natural selection when it operates between genotypes of different species (reproductively isolated groups). What does natural selection look like when it is occurring between Dandelions and Sequoia trees? Since Dandelions and Sequoia trees cannot exchange genes, is natural selection not irrelevant at the interspecific level?

It took a long time before I even asked myself this question but, once I took it seriously, it led me to the formal mathematical conclusion, given in Chapter 5, that the consequence of natural selection between genotypes of different species, far from being meaningless simply because it is largely irrelevant to evolution, is the key to understanding community assembly and helps to reconcile seemingly contradictory views of the nature of ecological communities that have coexisted uneasily for more than a century (Clements 1916, Braun-Blanquet 1919, Tansley 1920, Cooper 1926, Gleason 1926, Tansley 1935). The philosopher Daniel Dennett (1995) described natural selection as "Darwin's dangerous idea" because it is a process that is not limited to evolution or even to biology. Natural selection has the ability to leak out, infect, and conceptually modify non-evolutionary domains. This is because natural selection is not really a biological phenomenon at all. Natural selection is a general domain-neutral algorithm. Natural selection is a logic machine. This machine takes heritable variation of properties between replicators as input and forces these replicators, by the brute force of its axioms, to scale fitness landscapes in the "design space" of these properties. This is true whether or not[2] the replicators

---

[2] The efficiency with which natural selection can scale fitness landscapes is greater if the replicators can exchange such heritable information because this increases the variation in such properties.

exchange heritable information about their properties. "Evolution" is what happens when this logic machine forces interbreeding genotypes possessing heritable phenotypic variation to scale fitness landscapes. "Optimization of complex objective functions" is what happens when this logic machine forces replicating computer algorithms[3] to scale fitness landscapes (Munakata 1998). Finally, and most importantly for this book, "community assembly" is what happens when this logic machine forces reproductively isolated geno-types (i.e. individuals of different species) possessing heritable phenotypic variation to scale fitness landscapes.

In order to sketch out the development of the ideas in this book, I hope that you will excuse a conceit on my part so that I can describe my personal journey. I do this, not because mine is a particularly profound or even unusual one, but simply because the structure of the book reflects it and because the conclusions that I have reached are undoubtedly constrained by it.

My doctoral research began in 1983 on the topic of plant zonation along freshwater shorelines. In a river or marsh that experiences fluctuating water levels, when one goes from the area that is continuously inundated to the area that is never flooded, one typically sees a turnover of species (zonation) such that particular groups of species are consistently found at different levels of water depth. The motivating question for my thesis was as simple as it was naive: How do freshwater plant species assemble themselves into recognizable zones? It was a question concerning community assembly. I began, as seemed logical to me at the time, at the community level by studying the pattern of species' distributional boundaries as a function of water depth along the shoreline. I asked whether there really were distinct "communities" (vegetation zones) along this environmental gradient. This was, by 1983, already a classic question in plant community ecology that had occupied several generations of plant ecologists without resolution. I concluded (Shipley and Keddy 1987) that the question was unresolved because it was ill-posed with respect to its causal generating mechanisms. I took my conclusion seriously and decided that I should abandon such community-level research in favor of experimental field studies that

---

[3] Genetic algorithms form the basis of computer programs that are designed to solve a complex optimization problem. They do this by starting with an initial (even random) set of algorithms and compare the output of each algorithm to the value to be optimized (thus a "fitness" function). A subset are selected to "survive" with a probability proportional to their "fitness" and then (usually) the surviving algorithm codes are "mated" resulting in new hybrid "offspring" codes, along with some random "mutations"; this generates new variation. The process is iterated many times. Even very complex optimization problems, like the famous "travelling salesman" problem for which no exact solution is known, can be quickly and efficiently solved using this process of selection.

concentrated on the mechanisms themselves. I decided to descend to the
conceptual level of population biology.

I then embarked on a field experiment in which I planted individuals of
three dominant species, typical of different zones along the water-depth
gradient, both alone and in the presence of the natural surrounding vegetation
in each zone. The idea was to determine the relative importance of competitive vs. abiotic factors affecting growth and survival and how these factors
might generate vegetation zones. After two seasons of hard work I had an
answer (Shipley *et al.* 1991). Unfortunately, my answer had very unsettling
implications for me unless I was willing to ignore the original reason that
justified the study. The calculation was straightforward and depressing. If it
took two years for one ecologist to study competition involving three species
in one site then how long would it take to understand the zonation of
freshwater aquatic species in general growing in sites whose environmental
conditions were almost as unique as the species themselves? If the problem is
expanded to include all plant species along all environmental gradients then it
was clear that I faced a problem. In fact, plant community ecology as a
discipline faces the same problem using this research strategy since the
number of plant species is many orders of magnitude larger than the number
of plant ecologists. This point is the main topic of Chapter 2. Notice that, in
using this approach, I was implicitly viewing the plant community as a
machine with identifiable parts that could be isolated, manipulated, and then
put back together.

Being young, and being in need of an advanced degree so that I wouldn't
have to do real work, I didn't give up. I decided to descend yet another
conceptual level in order to tackle the question from the perspective of
functional plant ecology (Bradshaw 1987, Calow 1987). The logic behind this
third decision was as simple as for the previous ones. If there are too many
species interacting with too many environmental variables to hope to put
together a community piece by piece, then forget about species. The appeal of
plant traits is that there are many fewer important functional traits than there
are plant species. Individual plants migrate or fail to disperse, they grow or
die, they reproduce or fail to set seed, due to their morphological, physiological and phenological attributes in interaction with the surrounding
environment (both biotic and abiotic). It is not too much of a simplification to
say that traits (phenotypic properties) link genotypes to the environment.
Furthermore, plant traits are generalizable to a degree that plant species are
not. Knowing about the relative competitive abilities of *Acorus calamus* L.
versus *Carex crinita* Lam. at a given site does not allow one to predict the
competitive ability of any other two species at any other site. On the other

hand, if one can relate competitive ability (or any other ecological process) between species to some function of their traits, then one can potentially extend this to any pair of species possessing such traits. I therefore finished my doctoral thesis by studying the tradeoffs between a set of functional traits for a large selection of the plant species growing at my site (Shipley *et al.* 1989). One of the most important empirical results of functional plant ecology is that certain typical values of plant traits are systematically found in similar environmental contexts even though the taxonomic composition differs; this is part of the definition of a "plant strategy" (Grime 2007). I could see Grime's strategies emerge from the patterns of trait correlations in my study species.

After finishing my field work, and while writing up my results, I knew that I had still not answered the motivating question of the thesis since I had no way of going from plant traits to vegetation structure except in a very vague and qualitative way. I therefore brought the following problem to Len Lefkovitch, one of my thesis advisors and a very good statistician: If we know the typical values of a set of traits that should be found in a given environment (i.e. the plant strategy) and if we know the trait values of each species that could potentially reach the site, then how can we predict which species will actually be found, and at what abundances? There is no unique mathematical solution to this problem (it is under-identified) because there are more species than there are traits, but Len pointed out that this problem was similar to those studied in statistical mechanics, and suggested that I begin reading the work of a physicist, Edwin Jaynes, who had formulated the relationship between Information Theory and statistical mechanics through the Maximum Entropy Formalism (Jaynes 1957a, b, 1971). After reading Jaynes' papers I realized that this was both a potential way to solve my problem and a completely different way of viewing community assembly from the dominant one based on population dynamics and taxonomy. I wrote up these ideas relating community assembly and statistical mechanics via the Maximum Entropy Formalism as a fourth chapter of my thesis but was advised by my thesis committee to remove it since there were no empirical data to test the idea and because it was incomplete as an explanation.

I followed this advice (my thesis committee was correct on both counts) and the final version of the thesis (Shipley 1987b) did not contain the missing fourth chapter. Chapter 4 of this book contains the seeds of my lost thesis chapter. It has taken more than 20 years for these seeds to germinate and grow. The basic idea of applying Jaynes' Maximum Entropy Formalism to community assembly continued to feel right during the intervening years even though I couldn't completely articulate why, but one important stumbling block remained. The Maximum Entropy Formalism requires one to specify "constraints" on a process of random allocation (more correctly, on our information concerning this allocation) and I could not see how to formally and mathematically derive such constraints from ecological processes, as opposed to purely mathematical ones, in any consistent and general way. I put the problem to the back of my mind for many years although I continued to think about it occasionally, usually late at night when such vague ideas seem more promising.

It was only after collaborating with Éric Garnier and Denis Vile, and after stumbling by chance onto a paper describing a new and simple way of performing the calculations of the Maximum Entropy Formalism (Della Pietra *et al.* 1997), that I allowed myself to think about the problem during the day. Éric Garnier was interested in using "community-aggregated traits" to link plant traits to ecosystem processes. A community-aggregated trait is an average trait value in which the trait value of each species is weighted by the relative abundance of the species at the site. I realized that a community-aggregated trait was the same thing as the constraint equations of the Maximum Entropy Formalism and began to think about how population dynamics would lead to a community-aggregated trait. It was only when I removed my conceptual blinders[4] about natural selection at the interspecific level that I began to see the link between population dynamics, natural selection, community-aggregated traits, and community assembly.

I have now come to view the process of community assembly as a "statistical mechanistic" one. I place these words in quotations because I am referring only to the basic idea of a macroscopic pattern arising from random microstate allocations that are forced to obey macroscopic constraints. Classical statistical mechanics involves a number of important physical assumptions that are not appropriate to ecological communities (Haegeman and Loreau 2008, Shipley 2009). Chance allocations of resources at the level of microstates interact with natural selection between reproductively isolated genotypes. Natural selection biases such random allocations due to fitness differences between genotypes, leading to repeatable community structures at the macrostate level. The analogy that emerges is of Nature as an immense casino. The species play craps with loaded dice for resource payoffs. There is no guarantee of success, only a probability of success. The dice that each species uses are biased due to the unique traits that each possesses, thus

---

[4] An important paper in this regard was Van Hulst (Van Hulst 1992) who showed that the equations of dynamic game theory – a theory usually applied to genotypes of a single species – could be applied to model ecological succession by viewing each species as an asexual genotype.

weighting the probability one way or another, but whether or not the bias helps or hurts the species depends on the nature of the tables (environments) on which the dice are thrown. It is wrong to think that this view of community structure is simply an intermediate position along a conceptual axis whose limits are communities as "individualistic" assemblages and as deterministic "units". As Sewall Wright (1967) stated: *"The Darwinian process of continuous interplay of a random and a selective process is not intermediate between pure chance and pure determinism, but is in its consequences qualitatively utterly different from either."*

This book is therefore an exploration of these themes. I want the method that is developed in the book to be both applicable to real ecological communities in the field and be sound as an explanatory tool. Therefore, I hope that these ideas will be evaluated based both on their instrumental usefulness and on their explanatory unification. The greatest compliment that others can pay to these ideas will be to criticize them, expose their weaknesses, and then improve them by concentrating simultaneously on empirical application and theoretical development. Certainly, the theoretical development presented in this book as still incomplete. For instance, in this book the community-aggregated traits point downwards from communities to species but the origin of the idea of a community-aggregated trait points upwards towards ecosystems, not downwards towards species. The idea of a community-aggregated trait comes from the "biomass ratio" hypothesis put forward by Grime (1998) and proposes that "the extent to which the traits of species affect ecosystem properties is likely to be strongly related to the contributions of the species . . ." to community structure in terms of biomass abundance. Recent work has supported this hypothesis (Garnier *et al.* 2004, Kazakou *et al.* 2006, Vile *et al.* 2006a, Garnier *et al.* 2007, Violle *et al.* 2007b). Perhaps, once the ideas in this book have been properly tested and refined, they can be extended upwards to ecosystems; certainly there is still work to do in linking community assembly via traits to ecosystem properties.

2

Population-based models of community assembly

I approach this chapter with trepidation. Since the model proposed in this book is rather different from traditional models in community ecology, I want to discuss those aspects of model-building that will allow you to judge it relative to others. It might be fun to embark on a long-winded discussion about the philosophy of modeling in ecology, but only on a Friday afternoon, only if accompanied by cool beer and good friends, and especially if done in moderation. Since these conditions are not present (except, I hope, for the presence of friends) I will be brief.

Another reason why I am uneasy with this chapter is that I want to avoid giving the impression that I view previous modeling attempts in community ecology to be useless. On the contrary, these previous attempts were inspired (and have inspired me) and are of high quality. At the same time, it is important that I contrast my approach with previous ones. It is important that you understand both the differences with other models and also why I think that previous modeling approaches have failed, at least by the criteria that I am using. After all, if even *I* don't think that the approach is an advance over previous attempts then why should you invest your time in reading this book? I reconcile these two contradictory messages by appealing to the nature of scientific research. Despite the misconceptions of politicians and science administrators, science does not move in a straight line from problem to solution. No one can know beforehand, neither I nor anyone else, whether an idea will work or fail and it usually takes many years before enough empirical evidence has been collected to allow a decision to be taken. The French word for a "researcher" is "un chercheur", which simply mean a "searcher". I like the French word better because that's what we do; we search for answers with no guarantee of finding. The best that any of us can do is to propose our ideas, confront them with Nature, and see what works. What counts in the end is not the elegance of the idea but its success in describing the structure and dynamics of plant communities.[1]

Modeling goals

Since it is a truism that you will never arrive at your destination if you don't know where you want to go, the first obvious question must be: what are models for? Odenbaugh (2005) suggests five reasons.

(1) Models are used to explore possibilities.
(2) Models give scientists simplified means by which they can investigate more complex systems.
(3) Models provide scientists with conceptual frameworks.
(4) Models can be used to generate accurate predictions.
(5) Models can be used to generate explanations.

These five reasons essentially break down into two scientific activities, hypothesis generation and hypothesis testing, and two types of science, private and public science. Odenbaugh's first three reasons relate to hypothesis generation and this phase of model development is a private one. The last two reasons relate to hypothesis testing and these form the public phase of model development. The boundary between hypothesis generation and testing is usually blurred in practice. Although most working scientists continually move back and forth between the generation and the testing of hypotheses, it is still important to keep these two activities conceptually distinct.

The generation of scientific ideas is a profoundly personal one and the history of science shows that good hypotheses can be generated by just about any means, of which mathematical models are only one tool amongst many. After all, Archimedes apparently discovered the laws of hydrostatics after jumping into a bathtub full of water and Kekulé conceived of the ring structure of benzene after falling asleep before a fire and dreaming of snakes biting their tails (Beveridge 1957). If you want to use models to help you in generating hypotheses ("explore possibilities", "provide conceptual frameworks") then that is your business. Only you can judge the inspirational value of the model, and the criteria for inspirational value will differ for each of us. On the other hand the criteria used to judge the predictive and explanatory value of a model

---

[1] Thomas Henry Huxley famously said that a tragedy in science was the slaying of a beautiful theory by an ugly fact. Using the literary meaning of "tragedy" (as Huxley presumably did) this is true but, when using the colloquial meaning, a scientific tragedy occurs when the beauty of a theory succeeds in diverting our attention from the ugly fact.

are more objective and public. Whether a model helps us to predict and explain the assembly and structure of ecological communities is a public matter. It is the business of the entire scientific community.

Every ecological model, whether verbal or mathematical, whether empirical or deductive, consists of a logical transformation of input information (assumptions and initial conditions) into output conclusions. Every model has an "*IF + logical operations →THEN*" structure. The "*IF*" part of the structure consists of the assumptions and initial conditions of the model. The rules of logic, usually (but not necessarily) formalized in mathematic statements, are then applied to these input assumptions and initial conditions. The "*THEN*" part of the structure consists of the logical conclusions, which are tautological restatements of the "*IF*" part and the logical operations. Even the simplest model has this form.

Assuming that the internal logic of the model is correct, there are therefore two general questions that can be asked of a model. First, are the input assumptions reasonable and are the initial conditions correct? Some of these input assumptions can involve causal relationships between variables (Shipley 2000a), although the necessity (even the possibility) of this is denied by ecologists who subscribe to a logical positivist view of science (Peters 1991). Second, do the logical conclusions agree with Nature? Depending on the purpose of the model, its predictive ability can be judged based on the answers to either one or both of these questions. It is certainly possible for a model to make accurate output predictions even though the input assumptions are incomplete or even wrong. If the model does not claim explanatory ability – if, in other words, the output predictions are the only things of interest – then this is perfectly valid.

Explanation consists of demonstrating both that the input assumptions agree with Nature and that the internal logic of the model necessarily leads to conclusions that also agree with Nature. The general structure of an explanation is the following.

(1) *IF* (assumptions X_1, X_2 ...) are true *THEN* (output predictions Y_1, Y_2 ...) must also be true.
(2) I have shown that assumptions X_1, X_2... agree with the empirical data.
(3) I have shown that the output conditions Y_1, Y_2... agree with the empirical data.
(4) *THEREFORE* my model (i.e. assumptions + logical operations) explains the output conditions.

Thus, scientific explanation is, according to Hempel (1965), an argument claiming that the phenomenon to be explained was to be expected in virtue of certain explanatory facts. Explanation is therefore a stronger criterion than prediction for judging a model but explanation is not a *different* criterion. The traditional dichotomy between predictive and explanatory models (Peters 1991) depends on which side of the model you are comparing to Nature. Because of this Janus-like property of explanation, in which a judgment of agreement between claims and observations must be made at both ends of the logical argument, an evaluation of the explanatory ability of a model is not always obvious. It is vitally important to remember that the logical inference can only go one way. A common mistake is to assume that, since the output conclusions agree with Nature, therefore that this provides evidence in favor of the assumptions and the explanatory value of the model. It does not. Evidence for explanation can only go from assumptions to predictions.

Prediction, relative prediction, generality and predictive risk

Prediction refers to the difference between what Nature reveals and what our model claims. Such claims can relate to either the input assumptions or the output conclusions. In this sense I am using the word "prediction" in a more general way than is traditional in ecology. If, for instance, a model assumes a homogeneous distribution of resources then this is an empirical claim that can be compared to Nature and so is a "prediction" that should be true if the model claims to explain it. My use of the word "prediction" is also more general than normally understood because the predictions can (and ideally should) involve claims about outcomes following manipulations, not simply about new passive observations. On the other hand, predictions can be "free" or "constrained" and any empirical judgment concerning explanatory ability has to take this into account. For instance, imagine that I construct a global model of tree species richness whose qualitative output prediction is that there are less tree species in a boreal forest than in a tropical rain forest. Since I know which biome has more tree species per hectare even before I construct such a model, the agreement between this prediction and Nature cannot be counted as strong evidence in favor of it. After all, any model that got such a "prediction" wrong would have been stillborn before it even got into the public domain. In such a model the only useful empirical question can involve the validity of the input assumptions and so this refers to explanatory ability. On the other hand, a model that produces a correct quantitative prediction of tree species richness should (if I didn't know the answer beforehand) count as stronger evidence in its favor. The important criterion in judging the predictive ability of a model is by how much *better* we can predict Nature, given the model, relative to what we would know without the model. This relative predictive ability can be extended to comparing models themselves. The predictive worth of a model can be judged relative to another model by comparing by how much our predictive ability has been improved, perhaps weighted by the number of bits of input information we need to make the prediction (Akaike 1973, Burnham and Anderson 2004). Finally, it is important to consider the generality of the predictive ability of a model. Every model involves boundary conditions, even if they are implicit. The predictive ability of the model output should only be judged relative to these boundary conditions. However, the generality of the predictions is a reflection of the generality of the boundary conditions.

These considerations define the predictive risk of a model. Models that accept a greater predictive risk obtain more empirical support if their predictions prove correct (Popper 1980). Predictive risk increases as our prior knowledge decreases concerning the likelihood of the model claims (i.e. what we knew before considering the model). Because quantitative values can take more values than qualitative values, predictive risk increases as the claims concerning how Nature should behave become quantitative rather than qualitative. Predictive risk also increases as the generality of the claims increase. This is because more general claims allow a greater range of possible values and this decreases our prior knowledge about the correct value.

There is not, to my knowledge, any formal statistical method of judging overall predictive risk except in the simplest models, but the idea is to consider each input assumption and output prediction as a hypothesis. How well can we specify the distribution of possible values for the predictions of each hypothesis before considering the model? This is the prior distribution. How well can we specify the distribution of possible values for the predictions of each hypothesis after applying the model? This is the posterior distribution. The greater the overlap between the prior and posterior distributions (the more similar our knowledge before and after applying the model), the less that we can learn from the model, the less new information we gain even if the model is correct, and therefore the less predictive risk that is associated with the model.[2] This is because we either pretty much knew what to expect even before we proposed the model (the prior distribution was very narrow) or else because the model has made predictions that are no more precise than what we already knew.

---

[2] Predictive risk doesn't inform us about the truth of the model. After all, if we already knew exactly how a phenomenon arises and behaves before we propose our model then our model would risk very little but still be correct.

Judging explanatory ability is even less easy, partly because the level of explanatory ability that is appropriate has a strong subjective component. Nonetheless, explanatory ability consists of demonstrating (i) that the assumptions of the model actually apply to Nature (i.e. predictive ability with respect to assumptions), (ii) that the outputs of the model also apply to Nature (i.e. predictive ability with respect to the outputs) and (iii) that the assumptions logically entail ("account for") the outputs of the model. Each of these three parts must hold for a model to claim explanatory ability.[3] Since the evaluation of parts (i) and (iii) of an explanation is an evaluation of predictive ability (in the general way that I am using the word) and since part (ii) is not an empirical question but one of logic, then it is reasonable to talk about relative explanatory ability. The subjective component involves the nature and number of assumptions. All else being equal, we should prefer a simple explanation to a complex one if both have equal predictive ability; this is simply the application of Ockham's Razor.[4] However, every explanation (i.e. a logical argument) must begin somewhere; this is the "*IF*" part of the argument. By definition, the "*IF*" claims (the assumptions) are not themselves explained. Whether the assumptions begin at the right place is a personal judgment that depends on the detail of explanation that is sought.

Thus, relative predictive and explanatory ability are the criteria that I will use to evaluate different models.

Conceptual frameworks

As Odenbaugh (2005) has pointed out, models also provide conceptual frameworks. A conceptual framework is a way of "seeing" phenomena; that is, a way of concentrating on certain aspects of reality rather than others. I think that Odenbaugh has overstated the utility of models in providing conceptual frameworks. Most models don't provide conceptual frameworks so much as they make explicit pre-existing conceptual frameworks within which the modeler is

---

[3] Some people (Hempel 1965) view explanation from a purely logical vantage point and make no distinction between causes and effects. Realist philosophers add the requirement that explanation follow causal direction. For instance, imagine that you have three variables (air pressure, the presence of a storm and the height of mercury in a barometer) and derive the mathematical and logical relationships linking them in a model. Given this model, you can just as well derive the presence of the storm from the barometer as vice versa. For Hempel, both are perfectly acceptable logical explanations. For realists (and for me) the input assumptions of causality (the ordering of the variables in the model) must also agree with Nature for a model to be a valid explanation.

[4] "*Entia non sunt multiplicanda praetor necessitatum*"; "entities should not be multiplied beyond necessity".

Figure 2.1. A test of your conceptual framework. What do you see?

working. It is therefore important to carefully think about the conceptual framework that supports – and simultaneously constrains – a model. As an introduction to this section, take a look at Figure 2.1. What do you see?

Some people see a stylish young woman looking away from them and with her hair falling down over her neck onto her shoulder.[5] After staring at the picture for a few minutes most people will suddenly see the young woman disappear only to be replaced with an old woman looking off to the left,

[5] This version comes from an anonymous German postcard published in 1888. Another version, more well-known in North America, comes from W. E. Hill's 1915 cartoon "My Wife and My Mother-in-Law." (*Puck* **16**, 11, Nov. 1915).

complete with a pimple on her large hooked nose. Some people must even be shown where to look in order to do this.[6] For other people the order is reversed; the old woman is clearly visible and, only with effort, does the stylish young woman appear. We can eventually see the two women in the picture but cannot see both at the same time except with great difficulty.

This optical illusion has nothing to do with "reality" and everything to do with how your brain works. We naturally seek out patterns and impose order on the input of our senses based on past experience. When the input information is ambiguous we also tend to more naturally seek certain types of patterns than others; this is studied in the field of Gestalt psychology. Why people, especially men, more easily see the stylish young woman probably says something about male psychology but I won't speculate further. In any case, it makes no sense to ask if the picture is "really" a young woman or an old hag. It is neither. In "reality" the picture is simply a pattern of ink on a page.

Species or traits?

Now let's move from optical illusions to models of community assembly. What do you see when you stand in a forest? This question is not a Zen kaon. What do you *see*? If you are a biologist, and especially if you are an ecologist, you will probably see plants belonging to different species. Whenever I bring visiting ecologists to a forest in my region (southern Quebec, Canada) they will immediately ask me to give them species' names. They will note the dominance by *Acer saccharum* and *Fagus grandifolia* in more mature stands, the fact that *Populus tremuloides* is common in younger stands or that *Thuja occidentalis* takes over in wetter sites. In other words, they *see* a plant community as an assemblage of taxonomic units just as some people *see* the pattern of ink on the page in Figure 2.1 as being a picture of a young woman. This is how I see plant communities as well unless I consciously struggle against it in the same way that I must force myself to see the two women in Figure 2.1. Even people having no training in biology, once they reach the age of about 10, tend to see vegetation through the lens of folk taxonomy even if they can't recite Latin binomials. It just feels natural. It's just *obvious* that a plant community consists of species interacting with each other and with the environment. On the other hand, young children don't see species. They see plants with differently shaped leaves. They see big plants and small ones. They see plants whose stems are made of wood and plants whose stems

---

[6] The necklace of the young woman is the mouth of the old woman.

are made of soft tissues. They see plant communities consisting of plants possessing different traits. These two different ways of seeing (species/traits), these two different conceptual frameworks, are the ecological equivalent of the optical illusion in Figure 2.1.

Here we have two different conceptual frameworks of plant communities and this will affect how we construct models of community assembly. Do plant communities consist of plants grouped into species? If so, then the rules of community assembly will be based on demographic rules describing how individuals of different species interact with each other. Do plant communities consist of plants possessing different traits? If so then the rules of community assembly will be based on trait-based rules describing how the possession of different traits increases or decreases the chances of success of different plants. Of course, since individuals of different species also have different values of such traits, the results would be translated into patterns between species but, in a trait-based approach, the population dynamics of different species are a consequence of differences in traits, not a starting point. Since the goal of this book is to develop a mathematical model of plant communities based on traits, it is important to contrast this approach with the historically dominant approach[7] based on population biology.

Keddy (1990), while writing in Kyoto, Japan (home to one of the world's great Zen Buddhist temples), noted that one of the basic rules of Buddhist psychology is that people tend to become caught in habitual ways of interpreting reality and begin to confuse these habitual ways of seeing with reality itself. He suggested that the overwhelming tendency of professional biologists to "see" vegetation as plants distributed amongst species, and to assume that this is the only way of seeing plant communities, is a case of confusing the interpretation with the reality itself. This chapter explores the attempts of ecologists to understand and model community assembly from the majority view: communities composed of taxonomic entities (species). The next chapter will explore community assembly from the minority view of communities composed of plants possessing traits.

Let's follow the logical sequence of questions that are generated by viewing communities as collections of species. The first question would be the one posed (and answered) by Darwin and Wallace: where do these species

---

[7] Although what follows is necessarily a critique of the population-based approach to community assembly, I do not want you to get the impression that I am denigrating either the scientific quality or the importance of this approach. I also want to emphasize that the critique only applies to population biology *when extended to the community level*, not to population biology at the level of individual populations. In the end, scientific models can only be judged *post hoc* based on empirical evidence.

come from? This leads to the long tradition of evolutionary biology. Knowing how species arise, the second question might be: "why are there so many species?" (Hutchinson 1959). This leads to questions of taxonomic biodiversity. As soon as we ask why there are the observed number of species at a site, rather than more or less, then we are led to questions concerning the determinants of coexistence and exclusion of species. Finally, when we ask why we see species A, B and C rather than species X, Y and Z at a given site then we are asking about the rules governing the assembly of species over time and space. Note that each of these questions follows from the fact that we saw the vegetation as being composed of species; if we saw vegetation as being composed of plants possessing traits then we would pose very different questions (Keddy 1990). Because we saw the species as the fundamental unit of description, species become the state variables in our models.

Things or properties of things?

Besides species versus traits there is another, but related, dichotomy that defines different conceptual approaches to the study of plant communities. When you think about how plant communities are structured, do you see things (individual plants) that dynamically interact during community assembly or do you see things that express different properties? To explain the difference, imagine the following thought experiment. You take 10 fair dice and place a tiny weak magnet in the depression of the "1" face of each die. The magnets are much too weak to cause a die to stick to a vertical surface and only add a bit of extra force pulling the "1" face towards a metal surface. Now, place them in a metal box, shake the box and let the dice come to rest on the floor of the box. When you open the box you can either *see* the 10 "things" (the dice) or the 6 "properties" (the face values of the dice). Imagine conducting the experiment, opening the lid of the box, and looking inside it. Which would you notice first, the dice or the distribution of faces? What if you had instead placed 1000 such dice in the box? Would you still concentrate on the dice or on the distribution of faces? If you concentrate on the dice (the things) then you would ask for an explanation of why each die came to fall as it did. An explanation for the assembly of this collection of dice would consist of a dynamic explanation of how each die came to be where it is. If you concentrate on the faces (the properties of the things) then you would ask for an explanation of why the distribution of faces came to be expressed as observed. An explanation for the assembly of this collection of faces would consist of a causal explanation of how the properties of each face affected its chances of falling as it did.

Dice or faces? Things or properties of things? These conceptual frameworks will determine how you model the system. This distinction is primarily psychological rather than philosophical or scientific. The first explanation views the outcome of the system as a consequence of the dynamics of each die. Once you work out how each die behaved then you would simply add these together to describe the outcome of the experiment and the distribution of faces. The whole is a consequence of the behavior of the parts. The second explanation views the outcome of the system as a consequence of the properties of the faces. You would work out how the physical properties of each possible face of a die affect the chances of the face landing up irrespective of the dynamics of any particular die.

If we want the first explanation then we must describe the dynamics of each die as a consequence of how it changes its three-dimensional spatial orientation over time until it comes to rest. The changes in spatial orientation will depend on collisions between it and each of the other dice as well as the walls of the container. This will lead to a dynamic model, probably expressed as a system of differential or difference equations, describing the changes in the three spatial coordinates of each of the dice relative to the box at time t. Solving this system of equations at equilibrium (i.e. when all the dice come to rest) will tell us the spatial orientation of each die and, as a by-product, the relative abundance of faces.

If we want the second explanation then we must describe how the probability of a die showing face j is affected by its properties. We might proceed as follows. A perfectly fair die has no differentiating property that would bias a particular side and therefore each face is equally likely: thus $p_j = 1/6$. But, since our dice each have a magnet on side 1, this will increase the chances, by an amount δ, that a given die will land with this face against the floor of the metal box. Since side 6 is opposite side 1 on a die, the magnet would also increase the chances that a given die will land with face 6 pointing up. Since there are no physical biases of the other sides we have $p_1 = 1/6 - \delta, p_6 = 1/6 + \delta$ and $p_2 = p_3 = p_4 = p_5 = 1/6$. Solving for δ will give us the probability of each face landing up. Notice that this argument makes no reference to the dynamics of any of the dice. This could be done using engineering principles (for simple biases like our die) or using the method developed in Chapter 4. We would then reason that, once we have accounted for the physical causes biasing each face, the dice are otherwise equivalent (or "exchangeable") and therefore all other causes affecting the dynamics of a particular die cancel out and can be treated as random noise. This second type of explanation does not view the problem as one of explaining the behavior of things (dice) but of explaining the constraints imposed by the different properties of things.

Population-based models of plant community assembly

These two sets of dichotomous conceptual frameworks (species/traits and things/properties) can serve as a classification of the various mathematical models that have been proposed to explain and predict the assembly of plant communities. It makes no more sense to ask which conceptual framework is "right" than it is to ask if Figure 2.1 is "really" a drawing of a young woman or an old hag. On the other hand, it does make sense to ask which conceptual framework is more "useful" in attaining our goal. Our goal is to develop a mathematical model that explains and predicts the assembly of plant communities in the field and at various spatial scales. Chapters 4 and 5 develop a model from the conceptual framework of traits and properties. However, the historically dominant modeling tradition has worked within the conceptual framework of species and things.

If we view plant communities as composed of species interacting with each other over time then it is only natural to construct models of community assembly that start with species (population biology) and that extends the ideas and methods developed in population biology in order to put these species together into communities. Such an approach began in animal ecology but this research program in plant ecology was most explicitly promoted[8] by John Harper (1982). In a book with the wonderful title *The Plant Community as a Working Mechanism*, evoking the image of a plant community as a complex pocket watch composed of interacting parts, Harper gives his prescription for the best strategy to understand community assembly: "*[I]f we accept (again as an act of faith) that the activities of communities of organisms are no more that the sum of the activities of their parts plus their interactions it becomes appropriate to break down the whole into the parts and study them separately. Subsequently, it should, again as an act of faith, be possible to reassemble the whole, stage by stage, and approach an understanding of its workings*". This analogy of the plant community as a machine, with its attendant reductionist research program, has been the dominant one in the field of community assembly. In what follows I will briefly look at three of the most

---

[8] Harper actually argues against a purely taxonomic approach and for an approach based on population biology but this was only as a matter of which variables to measure.

influential mathematical models within this dominant conceptual framework with respect to the assembly of plant communities.

Lotka–Volterra models of community assembly

These models are so well known, and have been extended in so many ways by theoretical ecologists, that we won't spend too much time on them. They do, however, possess some important properties that are shared by all such "community-as-machine" models and so serve as useful pedagogical devices. A good historical description of the development of this class of models is given in Hutchinson (1978).

We start with the undeniable fact that all biological populations have the potential for exponential growth, expressed by the equation: $dn_i(t)/dt = n_i(t)r_i(t)$ or its finite difference version: $n_i(t+1) = n_i(t)(1 + r_i(t))$. In these equations $n_i(t)$ is the number of individuals of species i at time t and $r_i(t)$ is the average net number of new individuals added or subtracted from the population by each existing individual at time t (or during one time interval for the difference equation); this is the per capita growth rate. We then move to another undeniable fact, namely that no biological population can actually realize this potential for any extended time without exhausting the resources of our planet. There must therefore be some feedback process reducing the per capita growth rate when population sizes get too large. Viewing the community as a complex machine composed of interacting species, we would then describe the feedback relationship reducing $r_i(t)$ as being a function of these species: $\frac{1}{n_i(t)}\frac{dn_i(t)}{dt} = f(n_1, n_2, \ldots, n_s)$. This equation is reminiscent of the dice in the box when we concentrated on the dice (the things) rather than on the distribution of faces (the properties of things). Note that the state variables of this equation are the population sizes of each species and that the dependent variable is the proportional change in population size. The classical equation, sometimes justified by using the first term of a Taylor series expansion[9] for each species, has the general form of $\frac{1}{n_i(t)}\frac{dn_i(t)}{dt} = a_{i1} - a_{i2}n_i(t) - \cdots a_{is}n_s(t)$ of which the logistic equation emerges if there is only one species: $\frac{1}{n_i(t)}\frac{dn_i(t)}{dt} = r\frac{(K-an(t))}{K}$. Actually, "emerges" is the wrong word because the logistic equation for a single species was the inspiration and the community (multi-species) equation is simply the logical extension. The dynamics of community assembly is then described by the following system

---

[9] $f(n_i) = a_in_i(t) + \beta_in_i^2(t) + \cdots.$

of nonlinear equations where K_i is the "carrying capacity" of species i, r_i is the maximum (i.e. potential) per capita growth rate of species i, and a_{ij} is the amount by which an average individual of species j will reduce the per capita growth rate of species i from its potential ("competition coefficients"):

$$\frac{dn_i(t)}{dt} = n_1(t)r_1 \left(\frac{K_1 - a_{11}n_1(t) - \cdots - a_{1s}n_s(t)}{K_1} \right)$$

$$\frac{dn_2(t)}{dt} = n_2(t)r_2 \left(\frac{K_2 - a_{21}n_1(t) - \cdots - a_{2s}n_s(t)}{K_2} \right)$$

$$\cdots$$

$$\frac{dn_s(t)}{dt} = n_s(t)r_s \left(\frac{K_s - a_{s1}n_1(t) - \cdots - a_{ss}n_s(t)}{K_s} \right). \qquad \text{(Eqn. 2.1)}$$

The matrix of competition coefficients is called a "competition matrix". Needless to say, this model is an incredibly simplified (caricaturized?) version of what goes on in any real plant community and a great danger of all mathematical models is that so many of these simplifying assumptions are hidden. For instance, by expressing the interaction between species i and j as "$n_i(t) \bullet n_j(t)$" we are implicitly assuming that individuals come into contact at random and uniformly throughout geographical space. By ignoring higher order terms of the Taylor expansion we are assuming a simple linear relationship between population sizes and the reduction in the per capita growth rates from their potential maxima. By ignoring population sizes before time t in our equations we are implicitly assuming that history is irrelevant (if you know the current population sizes then past ones don't matter). By treating the competition coefficients as constants we are implicitly assuming that changes in the environment (besides the population sizes of the species) are irrelevant or non-existent. And so on. And so on.

But, one might argue, the advantage of making such assumptions is that we produce a simplified version that both captures the essential properties of the system while being applicable in practice. We will look at a carefully conducted case study of an even simpler system later (involving only a single species!) to see how well even the qualitative behavior is captured, but the model is certainly not applicable. Given a species pool of S species we must measure $S(S + 2)$ parameters (the S^2 competition coefficients as well as r and K for each species) and this is the minimum number obtained when assuming that these parameters are constant over time and space. Given that a typical hectare of grassland in my area can contain 70 species, and ignoring all those other species that might invade but be excluded before I see them, this means that I would have to measure 5040 parameters! In 1994 Paul Keddy and I attempted to bring together all those published empirical studies that measured such competition matrices of plant species (Keddy and Shipley 1989, Shipley 1993). The largest such matrix involved only 13 species (Goldsmith 1978) although none of the studies actually compared the dynamics of community assembly with those predicted by the Lotka–Volterra equations; doing so would have extended the length of such experiments by many more years.

Empirical research has shown that every one of the simplifying assumptions listed above is wrong. If the quality of the input assumptions is poor, what about the predictive ability of the output predictions, namely the actual population sizes over time or at equilibrium? Might this model be a good predictive device for output patterns even if it cannot properly explain them? Lotka (1925) published his model in 1925. Volterra (1926)[10] published the genesis of this model in 1926. Gause (1934) published his studies of laboratory populations of two species of *Paramecium* in culture bottles in 1934. Since then there have been a few other experiments, reviewed in Hutchinson (1978), that attempt to actually measure these parameters and compare the resulting dynamics to those predicted, including some studies by Clatworthy and Harper (1962) involving competition by three species of Duckweeds and *Salvinia natans* growing in laboratory conditions. Note that these simplified systems are those that would most likely accord with the restrictive input assumptions of the model. The most complete experiment of which I am aware involved five species of *Protozoa* (Vandermeer 1969), again in laboratory cultures. Vandermeer's study followed the dynamics of each species over time and showed qualitative, but not quantitative, agreement with the model predictions. The most complete test involving rooted plants was done by Roxburgh and Wilson (2000a, b). They chose a very simple system involving seven herbaceous species growing in a continually mown lawn and grew them in all pairwise combinations in order to obtain the competition matrix. Using the competition matrix they derived qualitative predictions of coexistence involving six three-species sets and the seven-species set. Five of the six three-species sets and the seven-species set showed significant deviations from predictions of coexistence (Dormann and Roxburgh 2005). They then applied various perturbations to the seven-species mixture and tested whether the community would return to its undisturbed state. They concluded that "*. . . whilst the theory may have biological relevance to at least simple communities, such as the lawn system used here, difficulties in its practical application, and more importantly, the restrictive assumptions on which the theory is based, will limit its relevance in*

[10] Translation by Wells in the appendix to Chapman (Chapman 1931).

most natural systems. These results call into question the generality of a large volume of theoretical studies based on these methods". That is the empirical basis of the Lotka–Volterra model of community assembly. The agreement between Nature and the input assumptions is poor. The agreement between Nature and the output predictions is also poor even though the empirical systems used were purposefully chosen for their simplicity and their agreement with the assumptions. If "explanation" means demonstrating that factually correct input assumptions logically entails factually correct output predictions, then this model fails both at prediction and at explanation.

It is surprising that so much ecological theory rests on such a flimsy empirical foundation but perhaps we are missing the point. Richard Levins, himself the author of many such community-as-machine models, has already made many of the criticisms that I have just given (Levins 1966) in his paper *"The strategy of model building in population biology"*. His recommendation, which has been largely followed, was the following: *"Since we are really concerned in the long run with qualitative rather than quantitative results (which are only important in testing hypotheses) we can resort to very flexible models, often graphical, which generally assume that functions are increasing or decreasing, convex or concave, greater or less than some value, instead of specifying the mathematical form of the equation. This means that the predictions we can make are also expressed as inequalities as between tropical and temperate species, insular versus continental faunas, patchy versus uniform environments, etc"*.

If Levins is right then we are in trouble. Community ecologists must forever accept that we cannot produce models that make quantitative predictions under field conditions. Even if you do not subscribe to quantitative predictive ability as a key property of a good model, and accept qualitative predictions as being just fine, the news is still not good because the qualitative "predictions" of such models are almost never real predictions at all. This is because the qualitative trends are almost always known before the model is constructed. Being able to successfully re-create such qualitative patterns is not a strong test of the model, it is a prerequisite for public viewing and so such models run very little risk of falsification.

Tilman's resource-ratio model

Tilman (1982) published the first mathematical model of community assembly that was explicitly developed for plants, although the genesis of the idea was developed by MacArthur (1972). This model corrected two important weaknesses of the Lotka–Volterra family of models. The first weakness is that, since we must minimally estimate $S(S + 2)$ coefficients in order to use a Lotka–Volterra type model involving S species, any but the simplest "communities" are beyond our empirical reach. If we were to add one extra species to the Lotka–Volterra model then we would have to estimate $(S + 2)$ extra parameters and the problem gets worse with each added species. The second weakness is that the numerical values of the competition coefficients in the community matrix will actually vary across environments rather than being constants in the model. If we want to use the Lotka–Volterra model in E different environments then we must actually estimate $ES(S + 2)$ parameters. Given 30 species and 12 different points along an environment gradient (a rather small number in most natural settings, and one that we will actually encounter in Chapter 6) we would have to estimate 14 976 parameters! Unless we can specify how the strengths of interspecific interactions change across environments then, even if we could solve the first problem, we would still sink beneath the weight of the model.

Tilman's model reduces the empirical demands of parameter estimation by replacing the competition coefficients by a mechanism of competition based on resource competition. In essence, the model extends the demographic approach to resources as well as species, where population growth is modeled using a Monod equation (Monod 1950). For the case of non-exchangeable limiting resources (R_j) the model states:

$$\frac{1}{N_i(t)}\frac{dN_i(t)}{dt} = \min\left(\frac{r_i R_j(t)}{R_j(t) + k_{ij}} - m_i\right)$$

$$\frac{dR_j(t)}{dt} = a_j(S_j - R_j(t)) - \sum_{i=1}^{n} \frac{\frac{dN_i(t)}{dt} + m_i N_i(t)}{b_i}. \qquad \text{(Eqn. 2.2a, b)}$$

Equation 2.2a describes the population dynamics of each of the i species as consequence of per capita growth (determined by the amount of the resource that is limiting) minus mortality. Equation 2.2b describes the dynamics of each of the j resources as a consequence of supply minus consumption. The total amount of resource j (quanta[11] of photosynthetically active radiation or soil nutrients) that is available in the environment is given by S_j. The $(S_j - R_j(t))$ term is the amount of the resource that has not yet been captured by living plants. However, since soil nutrients exist in both available and non-available forms, the unit rate at which the resource is being converted from

---

[11] Actually, one usually considers open space (i.e. space that is not yet covered by leaves) as the resource.

unavailable to available forms (a_j) must also be considered. The rate at which available forms of the resource is being supplied to the environment at time t is therefore $a_j(S_j - R_j(t))$. The rate at which available forms of the resource is being removed from the environment and placed in living biomass is given by the second term of Equation 2.2b. At equilibrium, i.e. when the derivatives of Equation 2.2 are zero, and considering a single limiting resource, the solution is $R^*_{ij} = k_{ijm_i}/(r_i - m_i)$ based on the limiting resource and this parameter (R^*_{ij}) is a property of each species in the given environmental context. Given a single limiting resource, the species with the lowest R^* will competitively displace all other species at equilibrium, independent of the initial population densities.[12] The maximum number of coexisting species cannot exceed the number of limiting resources. Since the number of limiting resources of plants in nature at any one place is generally small this means that spatial and temporal variation in the parameters relating to resource availability must be included in the model.

For each species we must know four things: (i) the maximum potential per capita growth rate (r_i), (ii) the resource availability of resource j at which growth reaches half of this maximum (k_i), (iii) the mortality rate (m_i) and (iv) the number of individuals produced per unit resource (b_i). The first three parameters involve measurements that can be made for most plant species[13] using hydroponic or similar culture techniques while the per capita mortality rate must be measured in the field and will be more difficult and variable over time and space. For each resource we must know two things: (i) the total amount of the resource (S) available in the environment for the resource and (ii) the rate constant for the resource supply (a_j). These two parameters are easily measured in aquatic laboratory systems involving phytoplankton but they are extremely difficult to measure in the field. The rate constant (a_j) is particularly difficult to measure because this involves all of the complicated chemical and biological processes of weathering, decomposition of soils and movements of water. Furthermore, such processes vary over time and space in very complicated ways, even at small spatial scales. Assuming no spatial or temporal variation in the parameters, we must therefore measure $S(3 + R) + 2R$ things, where R is the number of limiting resources. However, the model requires spatial or temporal variation in S_j and a_j and so the empirical demands will be $S(3 + R) + 2RX$ where X is the number of measurements required to properly quantify this variability.

---

[12] This is not true when there is more than one limiting resource.
[13] Although this requires the untested assumption that such parameters of individual plants correspond to population values. It is also likely that the parameters are size dependent such that individual-level measurements will change as the plant grows. Of course, this problem is avoided when dealing with phytoplankton.

This is clearly an improvement over the Lotka–Volterra equations. Given our hypothetical hectare of grassland with its 70 species, and assuming two limiting resources and no spatial or temporal variation in the measured parameters, we would only have to measure 354 parameters instead of 5040 with the Lotka–Volterra equations; of course, the absence of spatial or temporal variation in such parameters is a big assumption to make. Equally important, changing the environment (i.e. the resource supply rates and thus the potential productivity) only involves estimating two new values for each environment. Finally, adding a new species only requires us to estimate four new parameters, independently of how many species are already included in the model. Of course, if we have 70 species in our community then we must measure $2RX$ more parameters in order to capture the variability needed for the model to allow all 70 species to exist.

Miller *et al.* (2005) reviewed the empirical support for this model after approximately 20 years. It might come as some surprise that of the 1333 papers citing Tilman's (1982) book at the time, only 26 studies actually reported proper empirical *tests* of the theory, of which Miller *et al.* (2005) count 42 individual tests. Of the 42 tests, 32 tests are based on laboratory or chemostat experiments involving phytoplankton or zooplankton and only four were field experiments. The tests were almost always based on qualitative predictions and did not include the dynamics leading up to equilibrium. The general conclusion of Miller *et al.* (2005) was that the theory worked well in aquatic chemostat systems but not in terrestrial field systems.

One likely reason for both the paucity of terrestrial field tests and the poor predictive ability in such systems is the difficulty of measuring nutrient dynamics in real spatially heterogeneous soils. One exception which seems to have worked well in controlled field experiments, at least in the small number of species and sites in which it has been attempted, is the qualitative prediction of competitive exclusion or coexistence at equilibrium based on the R^* criterion. This is done by growing monocultures of each species and then measuring the concentration, rather than the supply rate, of the limiting nutrient in solution in the soil until a constant value is reached. Unlike supply rates, nutrient concentrations in soil solutions can be easily measured using lysimeters or other standard methods in soil science. On the other hand such measurements require a long time for the monocultures to reduce nutrient levels to constant values – 11 years in Dybzinski and Tilman (2007) – and one might still have to follow temporal dynamics of the nutrients over the growing season (Violle *et al.* 2007a). However, the outcomes of such competition experiments alone cannot be used to predict the relative abundance of mixed species in more natural field conditions since the predicted outcomes consist of competitive exclusion of all but a single, or perhaps a few, species (Dybzinski and Tilman 2007, Tilman 2007). Dybzinski and Tilman (2007) suggest that the discrepancy between the R^* predictions and actual relative abundances might lie in tradeoffs between competition and colonization. In other words, the equilibrium conditions under which R^* can predict competitive exclusion or coexistence are not applicable! All of these points make it too difficult to actually apply such models to most plant communities in a way that leads to quantitative predictions of relative abundance in the field.

The Lotka–Volterra type models derive population dynamics directly from interactions between species, via the competition matrix. Tilman's resource-ratio model derives population dynamics directly from interactions between species and resources. Species interact only indirectly via their effects on resource levels. Since the ability of a plant to capture resources is determined by its morphology and physiology, i.e. its traits, the next obvious step in the mechanistic reduction is to explain changes in resource levels via differences in plant traits.

This modeling reduction is described in Tilman's (1988) book in which the growth of plants possessing different combinations of traits are simulated along gradients of resource availability and density-independent mortality (disturbance). The major defect in this trait-based model is the large number of parameters (21), many rather difficult to measure, for each species plus the number (3) of environmental parameters for each point in space that must be measured in order to convert the input assumptions and parameter values into output predictions (Shipley and Peters 1991). A super computer was needed to perform these simulations. This trait-based model, like the resource-ratio model that was its inspiration, predicted competitive exclusion of all but a few species at each point on a gradient and therefore still requires spatial and temporal heterogeneity and this must be measured in the field if the model is applicable. This further increases the number of measurements. A simplified version of the model that ignored interspecific differences in physiological parameters like photosynthetic rate or nutrient uptake rates of roots was shown to be based on incorrect assumptions relating relative growth rates with biomass partitioning (Shipley and Peters 1990, Garnier 1991, Shipley 2006). The more complete model has never been tested, presumably because of the difficulty of measuring the large number of plant and environmental parameters included in it.

Despite the empirical problems of applying Tilman's trait-based model to most plant communities, it shares a number of similarities with the model proposed in this book. Although some of the input assumptions of Tilman (1988), involving morphology, physiology and plant growth rates, were wrong there is no reason why a modified version could not work in principle. Certainly, in order to obtain a model that can be applied in the field, it will be necessary to reduce to a minimum the number of parameters (i.e. trait and environmental values) that must be measured and, as much as possible, avoid parameters that are difficult to measure. These, however, are empirical questions and, as will be described in Chapter 3, there now exist large trait data bases for many plant traits and there is also a concerted effort to enlarge, consolidate and expand them.

The main problem that I see with the modeling approach taken in Tilman (1988), and the main difference with the model proposed in this book, is the requirement that the passage from traits to community structure require the intermediate step of population dynamics. Because, in this conceptual framework, community structure is obtained by "adding together" the dynamics of each species, the mathematical model will only predict the structure of plant communities in the field if it can correctly model the population dynamics of each species in the field. This, to me, is the Achilles Heel of this approach because real population dynamics in the field are incredibly complex. The dynamics of real populations are affected by nutrient supply rates that vary over a few centimeters of soil (Bell and Lechowicz 1994) and over a growing season, where a thousand different chance events control seed dispersal differently for each species, where lifespans, growth stages and generation times differ for each species, where temperatures and precipitation fluctuate unpredictably over time and space at many different scales, and where the population dynamics of the herbivores, parasites and pathogens of the plants are at least as complex as those of the plants themselves! Yet, correctly modeling all of these interacting variables is a prerequisite to modeling community assembly if we attempt to obtain community structure by adding together the dynamics of each species.

The problem is potentially even more complex. The view implicit in communities-as-machine models is that unpredictable fluctuations in population sizes over time are due to environmental variability (the external factors outlined above) and measurement errors. Thus, in the absence of environmental variability, populations will either equilibrate or else settle into regular periodic oscillations. This is why, after assuming that such external variability doesn't exist or else can be summarized by average values (for example, average per capita density-independent mortality rates) such models are evaluated at (or close to) their equilibria. However, it is becoming clear that population dynamics have the potential to display chaotic internal dynamics, meaning that even small differences in parameter values will result in completely different, and unpredictable, changes in the subsequent dynamics. This results in population sizes that seemingly fluctuate randomly over time even in the absence of any external random inputs. May (1974, 1976) showed that even the simplest deterministic model of population dynamics with feedback, the logistic equation in difference form, can display chaotic dynamics. Subsequent theoretical work has shown that chaos can be generated in models involving competition (Huisman and Weissing 1999, Huisman *et al.* 2006), predator–prey interactions (Gilpin 1979) and food web dynamics (Hastings and Powell 1991, Klebanoff and Hastings 1994). Because such deterministic chaotic dynamics looks like random population fluctuations, and because it is difficult to distinguish between the two in the lengths of typical empirical population time series, the question of deterministic chaos has mostly remained theoretical although Gassmann *et al.* (2005) have documented a number of cases in vegetation dynamics. Examples of chaotic dynamics in controlled laboratory systems involving aquatic mesocosms can be found in Beninca *et al.* (2008) and Becks *et al.* (2005). A series of carefully controlled manipulative experiments involving a very simple empirical system that will be described below have shown very clearly the presence of deterministic chaos. Before looking at this study, let me give you a simple example of chaotic dynamics.

To illustrate the problem, consider the logistic equation written as a difference equation: $N(t + 1) - N(t)(1 + r) \cdot \left(\frac{K - N(t)}{K} \right)$. There are no random variables in this equation and, given the values of r, K and $N(0)$, the subsequent population sizes, $N(t)$, over time are completely deterministic and predictable. The first row of Figure 2.2 shows, from left to right, the changes in population size over time when $r = 0.7$ (i.e. $1 + r = 1.7$), when $r = 0.71$ (i.e. $1 + r = 1.71$) and the relationship between the population sizes at each time when comparing these two dynamics. We see the classic form of the discrete version of the logistic equation in both simulations. The tiny change in the value of r between the two simulations ($\Delta r = 0.71 - 0.70 = 0.01$) is mirrored by the tiny change in the population sizes at the same time interval over time. In other words, a tiny change in the per capita growth rate translates into only a tiny change in the predicted population dynamics. The second row of Figure 2.2 shows the same thing except that the values of r are 1.7 versus 1.71. Notice, first, that the time series look completely random in both simulations even though there is absolutely no random component in the model. Exactly the same equation generates seemingly random, and certainly very complex, population changes over time. Even more importantly, the same tiny change in r (0.01) results in huge changes in the population sizes predicted at the same time interval. In other words, a tiny change in the per

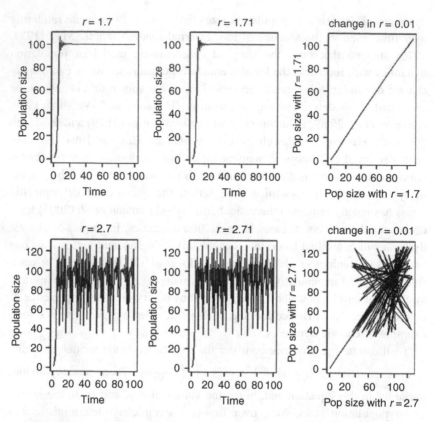

Figure 2.2. A simulation of a deterministic logistic difference equation, which models population change in a single species over time, showing chaotic and non-chaotic behavior in the same equation. Top row: Population sizes with a per capita growth rate of 0.7 (thus, $1 + r = 1.7$, first column), 0.71 (thus, $1 + r = 1.71$, second column) and the relationships between population sizes at the same time interval in the two simulations when the change in the per capita growth rate is 0.1. Bottom row: Population sizes with a per capita growth rate of 1.7 (thus, $1 + r = 2.7$ first column), 1.71 (thus, $1 + r = 2.71$, second column) and the relationships between population sizes at the same time interval in the two simulations when the change in the per capita growth rate is 0.1.

capita growth rate translates into very large, and seemingly random, differences in the population dynamics.

Think about what this would mean for our ability to produce accurate models of population dynamics if this mathematical property of chaotic dynamics exists in real population trajectories. If we make a mistake of 1% in estimating a per capita growth rate (1.71 instead of 1.70) then our model of population dynamics will be no more accurate than choosing random numbers! Yet, if r were measured in a real field population, a difference of 0.01 would be far less than the measurement error even if we were working with a huge sample size. Depending on the equation and the part of parameter space that is being explored, even changes in parameter values of 10^{-6} can produce equally spectacular differences; this is the famous "butterfly effect" (Gleick 1987).

So, are chaotic dynamics the exotic properties of mathematical equations, or do they exist in nature? Constantino *et al.* (2005) describe a series of very carefully conducted experiments designed to determine if such deterministic chaos could be found in a real biological population. The experimental system involved carefully controlled environmental conditions so that external stochastic factors could be removed, thus allowing dynamic chaos to emerge if it existed. Another important property of the experiments was that the system was incredibly simple; therefore, if chaotic dynamics occurred in this simple system, it increases the chances that it would occur in more complex natural situations.

The experimental population involved Flour Beetles, *Tribolium castaneum*, cultured in half-pint milk bottles that contain a constant amount of food (20 g of standard medium) placed in incubators and maintained at a constant temperature, relative humidity and amount of light. They derived a very simple model of the population dynamics of this species involving only three nonlinear difference equations representing the numbers of larvae, pupae and adults. The parameter values were (i) the average number of larvae produced per adult per time period (2 weeks), (ii) the average death rates per individual per time period, (iii) the amount of resources provided in each bottle and (iv) the cannibalism rates of adults and larvae. By exploring the dynamic behavior of these rather simple dynamic equations, given different combinations of the parameter values, and then experimentally manipulating this simple system to obtain the actual parameter values, they were able to demonstrate a host of different forms of complex dynamics. For instance, simply by changing the adult death rate, both the model and the actual population dynamics changed from a stable fixed point at equilibrium in which the population size remained constant over time, to a series of alternate cycles in which the population values cycled in a smooth manner, back to a stable fixed point, and then to a pattern of classic "period-doubling" (i.e. in which the number of cycles increased as a power function) leading to full chaotic dynamics. Different combinations of parameter values showed equally complex dynamics. The complexities of chaos theory were experimentally confirmed including, for particular combinations of parameters, extreme sensitivity to initial conditions.

This means that even small changes in parameter values quickly led to completely different dynamic trajectories. Finally, the authors showed how a combination of random external perturbations interacted with these intrinsically complex dynamics by pushing the system into different basins of attraction of the dynamic equations, further complicating the dynamics.

Now, if such complexities exist in such a simple laboratory system, with a single species growing in controlled conditions, imagine how complex it would be to correctly capture the dynamics of an entire species pool that is being assembled in the field when nutrient supply rates vary over a few centimeters of soil and over a growing season, where a thousand different chance events control seed dispersal differently for each species, where lifespans, growth stages and generation times differ for each species, where temperatures and precipitation fluctuate unpredictably from day to day and from year to year, and where the population dynamics of the herbivores, parasites and pathogens of the plants are just as complex as those of the plants themselves! Yet, this is a prerequisite to modeling community assembly by viewing a community as a collection of "things" in interaction.

Trying to predict the behavior of a plant community in this way is like trying to predict the behavior of a gas[14] by modeling the dynamics of each molecule as it smashes into each other molecule. No physicist has ever accomplished such a feat, not even through computer simulations, and the reason is similar. There are so many "things" interacting, and the dynamics of the system is so sensitive to small changes in parameter values, that any predictive ability is rapidly lost. Historically, the link between the properties of individual molecules and the behavior of gases ("communities" of molecules) was not forged through dynamics (Newtonian mechanics) but through statistical mechanics. The basic notion of statistical mechanics lies in the distinction between microstates (i.e. the spatial or temporal arrangement of molecules in different states, thus their individual dynamics) and macroscopic properties (i.e. average properties of the entire collection of molecules). The assumption is that the microscopic dynamics is sufficiently complex as to appear random but that certain constraints on the system (for instance, the fact that the total amount of energy in a gas in a thermodynamically closed system is constant) make some microstates more likely than others. These biased probabilities of microstate configurations are manifest as repeatable average properties (for instance temperatures, volumes and pressures) of the entire "community" of molecules. Knowing the values of these "community" averaged macroscopic properties allows us to then calculate the most likely

---

[14] Or a million dice inside a box. . .

microstates. We can do this by assuming that the microstate dynamics are sufficiently complex to be random but are constrained to agree with the macroscopic average properties. The relationships between chaotic systems and statistical mechanics are just beginning to be studied but some preliminary results are intriguing. For instance Egolf (2001) has demonstrated a simple spatially explicit system that displays chaotic dynamics but whose macroscopic structure is predicted using the same statistical mechanistic equations as given in Chapter 4.

This is the conceptual framework that I propose in this book. Combinations of functional traits interact with environmental gradients to determine the population dynamics of different genotypes. These different genotypes are differentially distributed among species. The population dynamics are much too complex to be predictable given the types of information at our disposal but different traits bias probabilities of reproduction, survival and dispersal (i.e. population dynamics) such that different environments favor different average trait values (i.e. natural selection). Because natural selection in a given environmental context (i.e. gradient position) constrains population dynamics and selects for repeatable average trait values, we can derive the most likely macrostate properties (i.e. species abundances) if we know these average trait values. In this sense, I am proposing a statistical mechanistic approach to community structure. However, the analogy should not be taken too far. Natural communities are no more like gases than they are like machines. The methodology that I will use, the Maximum Entropy Formalism, does not assume the restrictive mechanistic assumptions of classical statistical mechanics but rather is based on more general notions of Information Theory. None the less, the Maximum Entropy Formalism maintains the statistical mechanistic notions of microstate/macrostate properties and constrained randomness.

This is the topic of Chapter 4 but, before we get into the mathematical details, we should first think about what a trait-based approach to plant ecology would look like. This is the topic of the next chapter.

3

Trait-based community ecology

When you look at vegetation, you can concentrate on the *things* you see (plants or species) or on the *properties* of these things (the traits of plants). As the previous chapter has emphasized, most ecologists working at the population and community levels have concentrated on the things, not their properties, and, because of this conceptual framework, the process of community assembly has usually been seen as a demographic one. Despite this majority tradition there has always been, even from the beginnings of plant ecology as a scientific discipline, two minority approaches in the study of plant communities. The first minority approach, used especially by plant geographers, concentrated on morphological properties of plants and how, on average, such morphologies change as a function of major climatic variables such as temperature and precipitation (Warming and Vahl 1909, DuRietz 1931, Raunkiaer 1934, Holdridge 1947, Box 1981, 1995, 1996). In fact Warming, who can reasonably be called the father[1] of plant ecology, had an explicitly trait-based approach. A small group of American prairie ecologists applied this approach to the regional, rather than the global, scale during the first half of the twentieth century but without much impact on community ecology (Steiger 1930, DuRietz 1931, Dykserhuis 1949, Knight 1965). A second minority approach is based on the notion of ecological "strategies" (Southwood 1977). Grime's CSR scheme (Grime 1974, 1977) is perhaps best known to contemporary plant ecologists but Grime (2001, page 5) documents a continuous history dating back to MacLeod (1894). The underlying

---

[1] See Kurt, J. (2001). History of ecology. In: *Encyclopedia of Life Sciences*. John Wiley & Sons, Chichester. http://www.els.net/. "It was Eugenius Warming's *Lehrbuch der ökologischen Pflanzengeographie* that must be considered as the starting point of self-conscious ecology. This book was the first to use physiological relations between plants and their environment, and in addition biotic interactions to explain the moulding of the assemblages that plant geographers had described and classified, and it would set up a research agenda for decades to come."

assumption of this research program is that the incredible diversity of morphology, physiology and life history is constrained by natural selection and basic physical laws to vary along a small number of strategic dimensions. Furthermore, these strategic dimensions are assumed to covary with an equally small number of underlying environmental gradients.

If the trait-based approach to community assembly has existed from the beginnings of our science then why has it persisted for so long, like understory herbs on the forest floor, without ever replacing the dominant demographic approach? It has persisted, I would argue, because it is based on an observation that leads to an intuitively obvious conclusion. Since different species are associated with different environments then such common non-random patterns must be caused by some property of these different species. Demography (births, deaths, immigration and emigration) is simply the aggregate result of individual plants growing, reproducing, dispersing and dying. Since growth, reproduction, dispersal and survival are not caused by the Latin binomial with which taxonomists have baptized the species (i.e. looking at a plant as a thing), then what is left except for the morphological and physiological properties of the species? To deny this is to deny the existence of natural selection and to ignore the important contributions of physiological plant ecology.

If the persistence of the trait-based approach to community assembly is due to this obvious fact, then why has trait-based community ecology always remained a minority approach? The reason, I think, is because we have not yet figured out how to convert the intuitively obvious link between traits and environments into a logically rigorous and quantitative link between species and environments. In the end, community ecology deals with species, not traits. Tilman's model of plant strategies (Tilman 1988) comes close to forging an explicit logical link between traits, environments and species but, as I argued in the previous chapter, it is unlikely to succeed in any but the simplest and artificial cases. Grime's approach, based on plant strategies (Grime 1979, 2001), also comes close but lacks the explicit logical link that allows one to move from qualitative to quantitative predictions of abundance. We need some way of combining the empirical and theoretical strengths of each. I will propose one way of doing this in the next two chapters but we first need to look more carefully at how trait-based ecology works and how we might apply it to community assembly.

Trait-based ecology is associated with the subdisciplines of comparative plant ecology and functional plant ecology, the two being almost conceptual cognates. Most ecologists would recognize these as valid subdisciplines and yet it is rather difficult to determine what differentiates them from other types of plant ecology (Shipley 2007). Most subdisciplines of ecology ask questions relevant to a particular scale of organization. Physiological ecology studies individual plants or organs, population ecology studies populations of a single species, community ecology studies collections of species, ecosystem ecology studies fluxes of energy and matter in ecosystems, and so on. Comparative plant ecology spans all of these scales of biological organization. Other subdisciplines of ecology mostly measure types of variables that are specific to their subdiscipline. Physiological ecologists measure physiological and morphological attributes of plants, population biologists measure demographic attributes of populations, community ecologists measure abundances of species, and ecosystem ecologists measure fluxes and amounts of energy and mass. Comparative ecologists are so promiscuous that they measure all these variables (sometimes in the same study) and integration across temporal, spatial and organizational scales is often an objective of the research. The differentiating property of comparative/functional ecology is neither in the type of variables that are measured nor in the biological scale of organization in which such variables are measured, but rather in the way in which such variables are used. The unifying attributes of comparative plant ecology are (i) the use of organismal "traits" as explanatory variables, (ii) the comparison of such traits across many species and (iii) the implicit or explicit comparison of variation in these interspecific traits to environmental gradients. Comparative plant ecology is the study of interspecific relationships between organismal traits and environmental gradients.

Although a trait-based approach to community assembly will make heavy use of the notions and empirical results of comparative plant ecology, the purpose of this chapter is not to describe the field of comparative plant ecology or its empirical results. Such a description would require an entire book and there already exist much better books on this topic than I could write (Grime 2001, 2007). Rather, this chapter aims to clarify some key concepts and methods of analysis concerning traits, gradients, and how the two interact.

Traits

Functional ecologists have proposed and measured many traits that are believed to affect community structure and functioning (Knight 1965, Bazzaz 1979, Bazzaz and Sipe 1987, Gaudet and Keddy 1988, Lechowicz and Blais 1988, Keddy 1990, Westoby 1998, Weiher *et al.* 1999, Ackerly *et al.* 2000, Craine *et al.* 2002, Lavorel and Garnier 2002, Vendramini *et al.* 2002,

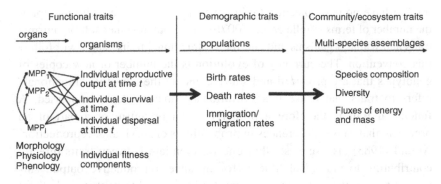

Figure 3.1. Summary of different types of traits existing at different levels of biological organization.

Cornelissen *et al.* 2004, Diaz *et al.* 2004, Garnier *et al.* 2004, Roche *et al.* 2004, Wright *et al.* 2004, Vile *et al.* 2006b, Cornelissen *et al.* 2007, Garnier *et al.* 2007, Violle *et al.* 2007b). So, if comparative plant ecology is the study of interspecific relationships between organismal traits and environmental gradients, then the next obvious question is: what is an organismal trait? The word "trait" is used in many different contexts and this could lead to confusion. Contrary to the views of some, a word cannot be forced to take a meaning simply by declaring[2] it to be so. I don't want to argue about how ecologists *should* use the word "trait" but only what I mean by the word. Figure 3.1 summarizes the notions and definitions. In this book I use a very broad definition: *a trait is any measurable property of a thing or an average property from a collection of things.* To me a trait is simply what a mathematician would call a "variable"; that is, a property possessed by all elements of a set but that can take different values in different elements. I will recognize four different types of "things" possessing traits and these are defined by levels of biological organization: organs,[3] individuals (organisms), populations and communities. An "organismal" trait is therefore a measurable property of an organism or an average property from a collection of organisms. Equivalent definitions follow for populational or community traits. Violle *et al.* (2007b) suggest that the terms "demographic parameter" and "community property" be used rather than "populational trait" or "community

---

[2] Quebec's *Office de la Langue Française* once stipulated that a "hot dog" was to be called "un chien chaud" in Quebec French. 40 years later, everyone in Quebec still calls it "un hot dog". Words take their meaning from usage, not pontification.

[3] One could also include cells or even organelles but few ecologists measure traits at this level.

trait". I have no real objection to this except that it unnecessarily multiplies the number of terms. Violle *et al.* (2007b) also group together sub-organismal traits and whole organisms and call these "functional traits" and I will follow this convention. The currency of evolution is the number of new copies of genotypes that are produced and survive and so the adjective "functional" refers to traits that cause this evolutionary function to be expressed. A functional trait is therefore a property of an organism or a part of an organism that causes differences in probabilities of survival or reproduction. Arnold (1983) also suggests that one differentiate between traits directly contributing to individual fitness (for instance, reproductive output, survival, lifespan, the ability to immigrate/emigrate) and traits that are indirect causes of individual fitness (i.e. morphology, physiology, phenology). I agree that this distinction is important because it makes explicit the direction of causality.

The hierarchical nature of traits suggests another important classification: levels of aggregation. I defined a trait as any measurable property of a thing or *an average property from a collection of things*. This qualification is needed because a trait can be defined at one level of organization but measured at a higher level. If a trait, measured at one level of biological organization, can be expressed as a (weighted) mean of values of the same trait at some lower level then the trait is said to be "aggregated" at the higher level. Consider specific leaf area (*SLA*), a trait usually defined as the projected surface area of a leaf divided by its dry mass (cm^2 g^{-1}). Since *SLA* is a property of a leaf it is, using my definition, a sub-organismal trait. Looking at Figure 3.2 (Violle *et al.* 2007b), you will see empirical results relating *SLA* (on the abscissa) to measures of net production efficiency (on the ordinate) when measured at three levels of biological organization (single leaves, whole plants or vegetation stands). Each point in Figure 3.2a, taken from Wright *et al.* (2004), shows the specific leaf area of a single leaf. Each point in Figure 3.2b, taken from Wright *et al.* (2001), shows the specific leaf area of a single plant. Each point in Figure 3.2c, taken from Garnier *et al.* (2004), is measured on all aboveground living vegetation in a series of small (0.5 m × 0.5 m) quadrats.

The data in Figure 3.2a cause no conceptual problem since *SLA* is measured at the same level as it is defined (single leaves). What about the data in Figure 3.2b? How should we interpret these *SLA* values when *SLA* is related to whole plants? A plant is not one big leaf; it can no more have a single *SLA* value than a population of people can have a single body weight. Notice that we could equivalently obtain the *SLA* values in Figure 3.2b in two different ways. First, we could measure the total projected surface area of all *k*

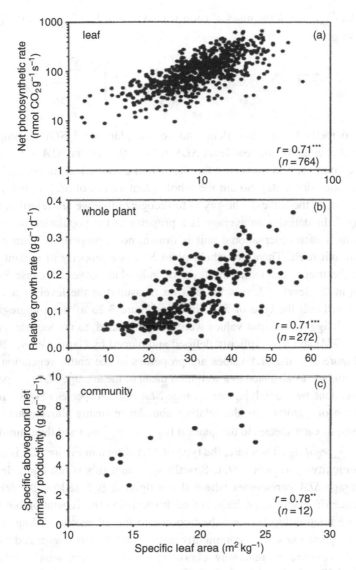

Figure 3.2. Relationships between specific leaf area and production efficiency measured at three levels of biological organization.

leaves of the plant ($\sum_k A_k$) and then divide by the total leaf dry mass ($\sum_k M_k$) of the plant; doing so gives $SLA = \sum_k A_k / \sum_k M_k$. Second, we could measure specific leaf area separately for each of the k leaves ($SLA_k = A_k/M_k$) and then calculate a weighted mean based on the relative contribution

$(r_k = M_k/\sum_k M_k)$ of the mass of each leaf to the total leaf mass of the plant; doing so gives

$$\overline{SLA} = \sum_k r_k SLA_k = \sum_k \left(\frac{M_k}{\sum_k M_k} \cdot \frac{A_k}{M_k} \right) = \frac{\sum_k A_k}{\sum_k M_k}.$$

The two methods are equivalent and so the plant-level *SLA* is simply a *weighted* average of the leaf-level *SLA* values; that is, the *SLA* value of an average "unit" of leaf mass (*not* the *SLA* of an average leaf unless each leaf had the same dry mass). So are the whole-plant values of *SLA* a property of the plant or of the leaves? The key is to recognize that we are referring to an "average". In statistics an average is a property of the population or sample from which each observational unit is drawn, not a property of each observational unit itself. Therefore whole-plant *SLA* is a property of a plant even though the trait is defined at the level of individual leaves. Because *SLA* is defined at the level of leaves but can be measured at the level of a whole plant, I will call the type of *SLA* used in Figure 3.2b a "plant-aggregated" *SLA*. Such aggregated trait values were first published, to my knowledge, by Grime (1974) and first explicitly defined and named by Garnier *et al.* (2004).

In Figure 3.2c the *SLA* values are properties of the entire vegetation of a quadrat and were calculated as a weighted mean of the average *SLA* per species. This was done by multiplying the average *SLA* of each species i ($\overline{SLA_i}$) in the vegetation of quadrat j by the relative abundance (using aboveground biomass, m_{ij}) of each species in the quadrat ($r_{ij} = m_{ij}/\sum_i m_{ij}$) and then summing: $SLA_j = \sum_i r_{ij}\overline{SLA_{ij}}$. Therefore, the type of *SLA* shown in Figure 3.2c is called a "community-aggregated" *SLA*. Strictly speaking, this is not equivalent to the average *SLA* per species where the weighting is based on the relative abundance (in mass) of the leaves of each species in the vegetation but there is a close relationship between the two due to the allometric scaling of leaf mass and plant mass. This community-aggregated *SLA* is interpreted as follows: if we were to randomly choose a series of very small "bits" of aboveground plant biomass in the quadrat, such that each bit was a pure sample of a given species, and associate the average *SLA* value of that species to each bit, then the *SLA* value of the average "bit" of biomass is the community-aggregated *SLA*.

We also have to distinguish between an aggregated trait and an analogous trait when moving between organizational levels. This distinction can be seen by looking at the ordinate of Figure 3.2 which shows the values of "net production efficiency", having units of $g\,g^{-1}\,s^{-1}$ or transformations of this; let's call this variable "X". Net production efficiency (X) is equivalent to a compound interest rate and is the net amount of new mass produced (grams or micromoles) by each existing amount of mass per unit of time (thus seconds or days). We could measure X at the level of an individual leaf (grams of new mass produced per dry mass of the leaf per second) and call it "net photosynthetic rate". We could measure X at the level of an individual plant (grams of new mass produced per dry mass of the plant per second) and call it "relative growth rate". We could measure X at the level of a quadrat of vegetation (grams of new aboveground mass produced per dry mass of aboveground mass of the vegetation per second) and call it "specific aboveground net primary productivity". In each case we are measuring the same underlying "thing" (net production efficiency) and it is clear that X is somehow related over these levels of organization. However, the definition of X (our "trait") is not strictly equivalent over these levels. In Figure 3.2a we are measuring the proportional increase in leaf mass over time,[4] in Figure 3.2b we are measuring the proportional increase in plant mass over time and in Figure 3.2c we are measuring the proportional increase in aboveground vegetation mass over time. However, plant relative growth rate is not a plant-aggregated value of leaf net photosynthetic rate because it is not simply a weighted average of leaf net photosynthetic rate. When we move from leaves to whole plants we must take into account mass loss through respiration of stems, roots and inflorescences and this mass is ignored in the leaf-level variable. Similarly, specific aboveground primary production cannot be calculated simply as a weighted average of whole-plant relative growth rate because, at the vegetation level, we have ignored belowground biomass. Net photosynthetic rates of leaves, relative growth rates of individual plants and specific aboveground primary production are "analogous" traits since they measure the same underlying process at different scales but are not simply related by aggregation. Note that if Garnier *et al.* (2004) had measured specific *total* primary net productivity rather than *aboveground* net primary productivity then this would be an aggregated mean of whole plant relative growth rate.[5]

A final distinction that some authors make is between a "response" trait and an "effect" trait (Diaz and Cabido 2001, Lavorel and Garnier 2002, Diaz *et al.* 2004). An effect trait is a functional trait, thus one defined at the

---

[4] Not really, since the new mass (carbon) that is produced is then partially exported out of the leaf.

[5] Since the RGR values in Figure 3.2b are maximum values, obtained using plants grown under optimal conditions in a hydroponic solution, these are not numerically equivalent to RGR values of the species in the field. However, there was a tight correlation between these maximum values and the community-aggregated RGR values measured in the field.

organismal or sub-organismal level, which causes changes in some ecosystem process. The functional trait therefore *affects* the ecosystem process. A response trait is one that does not have this property. For example differences in specific leaf area, a leaf-level trait, cause differences in leaf decomposition rate once the leaf has died (Cornelissen 1996, Cornelissen and Thompson 1997). This can be scaled up to the ecosystem level as a community-aggregated *SLA* where it causes differences in mixed-species litter decomposition rate and therefore nutrient dynamics in the soil (Garnier *et al.* 2004, Kazakou *et al.* 2006). Therefore, *SLA* is an effect trait. On the other hand, differences in seed size have no known consequences, when scaled up to the level of an ecosystem, on ecosystem properties. Therefore, seed size is a response trait.

An ecological result that cannot be generalized to new places or times cannot make a prediction and so cannot be the basis for a falsifiable model. This is an important limitation of any type of community ecology whose state variables are species. Such species-based studies can describe community composition but cannot predict community composition in new sites except at very local geographical and temporal scales. For instance a multivariate ordination of vegetation done in Australia could not possibly be applied to vegetation in boreal North America since the two regions have virtually no species in common. On the other hand, a relationship between average specific leaf area, temperature and precipitation that is developed in Australia can easily be applied to Canada's boreal forest. Whether the relationship actually has predictive success is an empirical question but this is quite different from species-based vegetation description where even the possibility is removed. Trait-based community ecology can be generalized since traits are properties possessed by all (or most) plants irrespective of geography or taxonomy.

However, for trait-based ecology to be useful in practice, we need to compile data bases of important traits for large numbers of traits. There is now a concerted effort to construct such data bases and to cross-reference them to geographically identified data bases of vegetation samples and environmental variables (soils, temperature and precipitation). For instance Moles *et al.* (2005) have compiled worldwide information on seed size for approximately 13 000 species. Wright *et al.* (2004) have compiled worldwide information of certain leaf traits (lifespan, maximum net photosynthetic rate, specific leaf area, leaf water content, respiration rate) for about 2500 species. Grime *et al.* (2007) give information on up to 21 traits for 822 species of British species. The Australian CRC gives certain trait information on a large number of Australian species (http://www.weeds.crc.org.au/projects/project_3_2_3_1. html). Fumito Koike (http://vege1.kan.ynu.ac.jp/traits/PlantTraitAsia.pdf) has compiled a trait data base of East and Southeast Asian plants.

Kuhn *et al.* (2004) have produced a trait data base for about 3500 German species. Kleyer (1995) has also published a trait data base of German plants (http://www.uni-oldenburg.de/landeco/21337.html). The European SynBioSys data set (http://www.synbiosys.alterra.nl/eu/) has vegetation information on 275 000 species records. Europeans are also constructing the LEDA data base (Knevel *et al.* 2003) of 26 life history traits (http://www.leda-traitbase. org/LEDAportal/). A research group based at the Centre d'écologie fonctionnelle et évolutive (CEFE, CNRS) in France have been compiling a worldwide trait data base (PLANTRAITS) involving up to 61 individual and populational traits. In the United States researchers are currently setting up a trait-based data set called TraitNet (http://www.columbia.edu/cu/traitnet/). These efforts at compiling and cross-referencing such trait and vegetation data bases are key to a trait-based approach to community assembly.

Habitat descriptors

Ecologists commonly describe certain habitat properties of species. We speak about "aquatic" species, "desert" species, "understory" species and so on. Clearly, such habitat descriptors are not traits as I have defined them since the descriptor (aquatic, desert, understory) is a property of the environment in which a species occurs, not a property of the plant itself. Perhaps the most complete system of habitat descriptors is Ellenberg's system of indicator values[6] (Ellenberg 1978). Such indicator values have been compiled for over 2700 species in central Europe and consist of ordinal ranks (1–9) that categorize the level of various environmental variables at which each species reaches its maximum abundance (light, temperature, continentality, moisture, soil pH, nutrients, salinity, heavy metal resistance). A number of studies have shown the generally good agreement between these ordinal ranks and actual environmental measurements, so long as the ordinal nature of the variables is taken into account (Thompson *et al.* 1993a, Schaffers and Sykora 2000, Diekmann 2003).

One can use such habitat descriptors (once translated into ordinal numbers) in the same way that one can use traits in the mathematical model that will be developed in the next chapter. The disadvantage of habitat descriptors is that they do not have explanatory value. Saying that a species is an "aquatic" species is a description of its ecological behavior, not an explanation for it. In terms of explanatory value, it is better to refer to those morphological and

---

[6] http://www.sci.muni.cz/botany/juice/mani.htm

physiological properties of plants that allow – or prevent – them from completing their life cycle under water because it links distribution to natural selection but, in the mathematical model that will be developed, the result would be the same in terms of predictive ability. The advantage of habitat descriptors is that they can provide predictive ability with less effort and can replace an entire set of functional traits that may be difficult to measure. Habitat descriptors can also be useful as very coarse "filters" when defining species pools; this will be discussed later in this chapter.

Environmental gradients

If comparative plant ecology is the study of interspecific relationships between organismal traits and environmental gradients, and if we are attempting a trait-based approach to community assembly, the next obvious question is: what is an environmental gradient? This may seem like a strange question because the notion of a gradient is firmly embedded within community ecology. None the less, it is important to be explicit concerning what I mean (and don't mean) by an environmental gradient, about the different types of environmental gradients that can be recognized, and how to measure them. In writing this book I looked for an explicit definition of an "environmental gradient" in ecology and, surprisingly, I could not find one. This contrasts with the term "gradient analysis" for which there are many definitions. For instance Leps and Smilauer (2003, page 25) define gradient analysis as "any method attempting to relate species composition to the (measured or hypothetical) environmental gradients". However, the closest they come to defining an "environmental gradient" is when they say that environmental gradients of soils ". . . are described (more or less) by the measured soil properties" (Leps and Smilauer 2003, page 5). Judging by the way ecologists generally use the word "gradient" one might think that environmental "gradients" and environmental "variables" are equivalent terms. However, judging by the way environmental gradients are measured, there is an important distinction between an environmental variable and an environmental gradient.

If we choose a series of spatial or temporal coordinates and measure the properties of the things that we find at each location, then we can conceptually divide these properties into those relating to the plants and those properties relating to everything else (i.e. the "environment"). As a first attempt, perhaps we could define an environmental gradient as any combination of such environmental properties? This first attempt is surely defective. One cannot choose any environmental variable or combination of environmental variables and call it an environmental "gradient" as the term is used in ecology. Community ecologists regularly produce ordinations of species composition and then look for particular combinations of environmental variables (which are then deemed to be environmental "gradients") that best account for the major trends in species composition; this is a fundamental goal of vegetation gradient analysis (Austin *et al.* 1984, Austin 1985, Minchin 1987, Ter Braak and Prentice 1988, Austin and Smith 1989, Braak 1994, Legendre and Legendre 1998). Since the combinations of abiotic variables to which vegetation ordination axes are related, and that define environmental gradients, are chosen specifically to maximize the amount of variance (or other measures of statistical fit) in species composition, then some combinations of the environmental variables will form an environmental gradient but others won't. As another example, ecologists regularly speak about environmental gradients of water availability, productivity, stress and disturbance but these are not attributes of the "environment" without reference to what the plants are doing. A closer look at how Grime (2001) defined "stress" and "disturbance" gradients leads to a precision. For instance, stress "... *consists of the phenomena which restrict photosynthetic production such as shortages of light, water, and mineral nutrients, or sub-optimal temperatures*" (Grime 2001, page 7). This definition is rather vague but Grime cannot mean that a stress gradient is simply a *list* of these abiotic variables (amounts of light, water, mineral nutrients, temperatures, etc.) but rather a specific combination of these variables that together cause – and therefore predict – the degree of photosynthetic production.[7] "Disturbance" is defined as "... *the partial or total destruction of the plant biomass and arises from the activities of herbivores, pathogens, man (trampling, mowing, and ploughing) and from phenomena such as wind-damage, frosting, droughting, soil erosion, and fire*". Again, he is not simply referring to a list of these phenomena. Disturbance takes different values and these values are determined by a particular combination of these phenomena that is linked to the amount of biomass that is destroyed. In each case an environmental gradient is characterized by a *relationship* between the environmental variables and some plant property.

More generally we might say that if Y is some property of the vegetation and $\mathbf{e} = \{e_1, e_2, \ldots, e_j\}$ are a list of environmental variables then an "environmental gradient" in ecology is not the list of environmental variables ($\mathbf{e}$) but rather the function, $g(\mathbf{e})$, that specifies the causal link between the

---

[7] I will leave aside the complication with this definition that the same set of environmental conditions will produce different amounts of photosynthetic production depending on which plants are doing the producing.

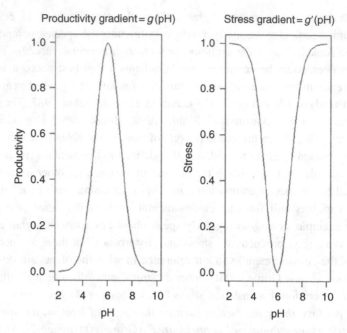

Figure 3.3. A hypothetical environmental gradient describing the relationship between a plant property (productivity) and an environmental variable (pH).

environmental variables and the plant property . Certainly, few direct gradient analyses actually test for a causal link but the assumption is implicit. For instance, in Figure 3.3 (left) the environmental *variable* is pH, the plant variable is productivity and the environmental *gradient* (the productivity gradient) is the function linking the two:

$$productivity(pH) = \frac{1}{0.4\sqrt{2\pi}}e^{\frac{-(pH-6)^2}{2}}.$$

We could translate our hypothetical productivity gradient into a stress gradient (Figure 3.3, right), using Grime's definition, as follows:

$$stress(pH) = 1 - \frac{1}{0.4\sqrt{2\pi}}e^{\frac{-(pH-6)^2}{2}}.$$

The best definition that I can propose for the ecological notion of an environmental gradient is: *a mathematical function, g(e), which maps values of a set of environmental variables onto some property of plants and (preferably) reflects the causal relationships between the environmental variables and the plant property.* Of course, many ecological studies refer to environmental gradients without writing down mathematical equations but, I would argue, such a functional mapping (in the mathematical sense) is implicit. What ecologists call an environmental *gradient* should more properly be called a *function* (in the mathematical sense). In fact, I would not use the word "gradient" at all if this word was not already so strongly embedded in ecology since the pre-existing mathematical definition of a gradient is quite different. Rather than "environmental gradients", a better term would be "vegetation–environment functions" but there is no sense[8] in using a word that no one else will use. In mathematics a gradient is a vector of partial slopes. If $f(x, y, z)$ is the function then the mathematical gradient of this function is $\nabla f = \left(\frac{\partial f}{\partial x}, \frac{\partial f}{\partial y}, \frac{\partial f}{\partial z}\right)$ and this is obviously not what ecologists mean by a "gradient". I don't know who introduced the term "gradient" into ecology. I am no historian, but the term did not seem to be part of the working vocabulary of some of the most influential plant ecologists of the first part of the twentieth century (Tansley 1920, Cooper 1926, Gleason 1926, Tansley 1935, Clements 1936). I can first trace the word to Whittaker, who used the word "gradient" is his classic studies of changes in species composition in the Great Smoky Mountains (Whittaker 1951, 1956, 1967). In his 1970 textbook (Whittaker 1970) he begins the section entitled "Species along environmental gradients" as follows: *"Consider first a single environmental gradient, which could be a long, even, uninterrupted slope of a mountain"*. In other words, Whittaker's transects really did describe gradients (i.e. rates of change or slopes) since his transects consisted of slopes up mountainsides!

Types of environmental gradients

Community ecologists distinguish between direct and indirect gradients and between simple and complex gradients. How do these relate to the definition I have just proposed? My definition refers to a direct gradient. The functions shown in Figure 3.3 would be simple direct gradients, judging from common usage of the term, because they are each a function of a single environmental variable (pH). Note that "simple" does not refer to the simplicity of the function but rather to the number of environmental variables in the function; perhaps we could call it a simple nonlinear environmental gradient. A

---

[8] 'When I use a word,' Humpty Dumpty said, in a rather scornful tone, 'it means just what I choose it to mean, neither more nor less.' 'The question is,' said Alice, 'whether you *can* make words mean so many different things.' 'The question is,' said Humpty Dumpty, 'which is to be master – that's all.' *Through the Looking Glass*, by Lewis Carroll.

complex environmental gradient is a function involving more than one environmental variable.

What about an "indirect" environmental gradient (Whittaker 1967)? Let's go back to the scenario that started this section: We choose a series of spatial coordinates and, at each coordinate, we record the abundance of each species of plant as well as the values of a set of environmental variables. If we measure the degree of similarity in the species composition of the vegetation (or some other property of the vegetation, like a set of community-aggregated traits) between each pair of spatial coordinates, and then order these spatial coordinates in such a way that the distance between coordinates reflects, as closely as possible, the distance between coordinates in terms of vegetational similarity, then the resulting ordering of coordinates (sites) is what community ecologists call an "indirect" gradient analysis. For mathematical reasons one actually works with degrees of dissimilarity (or distance). The indirect "gradients" produced by this procedure are mathematical functions that do not involve environmental variables at all and so are not really gradients as I have defined them. However, if one assumes that the composition of the vegetation at each site is caused by the environmental conditions of each site then the ordering of spatial coordinates that best describes the patterns of vegetation similarity (or trait similarity) must be the same ordering of spatial coordinates that best describes the similarity in the unknown environmental gradients that are being sought.

In geometry, a distance is a function between two points that satisfies three conditions: (i) the distance between a point and itself is zero, $d(x, x) = 0$, and the distance between a point and any other point is a positive number, $d(x, y) \geq 0$, (ii) the distance between points x and y is the same whether you go from x to y or vice versa, $d(x, y) = d(y, x)$ and (iii) the distance between two points is the shortest path between them, resulting in the "triangle inequality" between any set of three points: $d(x, z) \leq d(x, y) + d(y, z)$ (Figure 3.4). If the triangle *equality* holds for all points, i.e. $d(x, z) = d(x, y) + d(y, z)$, then we can arrange the points along a single axis (Figure 3.4a). If not, if $d(x, z) < d(x, y) + d(y, z)$, then we require more than one axis (Figure 3.4b). The process of ordering the spatial coordinates along axes of distance is called an ordination.

If we ignore the second (Y) axis in Figure 3.4b, and use only the predicted distances of our three spatial coordinates on the first axis, then we have "projected" our actual distances onto this axis (the open circle) and this projected distance will result in some error. If we have recorded our vegetational composition at n spatial coordinates then there will be a maximum of n axes. The length (or "importance") of each axis is measured as the maximum distance between points along it, or some mathematical transformation

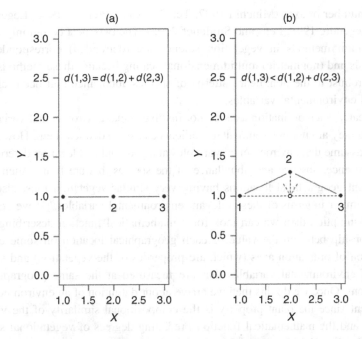

Figure 3.4. The location of three points on two axes in which the triangle equality holds (a), showing that the points lie on a single axis, and in which the triangle inequality holds (b) showing that at least two axes are necessary. The open circle in (b) shows the position of the point projected onto the first axis.

of this. If the length of an axis is small relative to others, for instance axis Y relative to axis X in Figure 3.4b, then ignoring it will not distort the true compositional distances too much. Why would we want to do this? First, because we might assume (or hope) that there are only a few unknown underlying environmental gradients and so the remaining axes represent random sampling variation, which can be ignored. Second, because indirect gradient analysis is really a set of exploratory methods whose purpose is to help us visualize major vegetational trends and develop hypotheses concerning what environmental gradients might be at work. Statistical methods that order spatial coordinates that are described by vegetational composition in this way are called ordination methods.

There is a rich literature describing different ways of measuring compositional distance and describing different ways of ordering the spatial coordinates in such a way as to maximize the agreement between measured distances (i.e. degrees of dissimilarity in species composition between spatial coordinates) and the distances predicted using some small (usually one or two) number of axes (Minchin 1987, Ter Braak and Prentice 1988, Legendre and Legendre 1998, Leps and Smilauer 2003). The two most commonly used ordination methods in vegetation science are (detrended) correspondence analysis and (nonmetric) multidimensional scaling because these methods are most robust to the nonlinear patterns of species abundance that occur along actual environmental variables.

Clearly, such ordination axes do not involve actual environmental variables and so they are not environmental gradients as I have defined them. However, if we assume that environmental variables at a geographical location determine the presence, absence and abundance of the species that are found, then two different geographical locations having very similar vegetation must also be very similar in terms of these relevant environmental variables.[9] If we accept this assumption then we can look for a mathematical function describing the relationship between the value of each geographical location on some combination of ordination axes (which are properties of the vegetation) and some set of environmental variables that are measured at the same geographical locations. Once we do this then we arrive at our definition of an environmental gradient since the plant property is the compositional similarity of the vegetation and the mathematical function predicting degrees of vegetational similarity, almost always a linear function, is based on environmental variables.

Before moving on I want to point out the confusion that often arises between environmental gradients, as I have defined them, and spatial gradients. There is a long tradition in community ecology of running transects up mountainsides (Whittaker 1956), perpendicular to shorelines (Keddy 1982, 1984, Shipley and Keddy 1987) and so on. Each physical location along such transects will have environmental variables that presumably have causal effects on plant fitness (temperature differences up the mountain, degrees of soil anoxia or water depth up the shoreline) as well as environmental variables that do not (altitude up the mountain, distance from the shoreline in the wetland or, in general, the spatial location itself). It is therefore important to carefully choose the relevant environmental variables when defining the gradient. The altitude on a mountain or the distance to a shoreline does not determine the abundance of a species or any other plant property and therefore one would not expect such a spatial gradient to have any ability to predict the plant property in a new location. On the other hand, temperature at

---

[9] The converse is not necessarily true. Two quadrats having equally different vegetational composition from some third quadrat might be equally different for very different reasons. There is only one way to have the same vegetational composition (the same species in the same abundances) but lots of different ways to have the same degree of different composition. This is the origin of the so-called "arch" effect in ordination diagrams.

particular altitudes up a mountain, or water depth along a shoreline, might affect the growth, survival or reproduction of plants and so would be appropriate environmental variables with which to define environmental gradients. The only reason to use (say) altitude up a mountain rather than other environmental variables is that altitude is easy to measure and also correlated with the more difficult to measure but causally efficacious environmental variables. However, the correlations between purely spatial coordinates and causally efficacious variables will not be stable when changing locations and this limits the generality of the environmental gradient. Environmental gradients should ideally be defined in the mathematical space of causally efficacious environmental variables rather like the n-dimensional niche space of Hutchinson (1978), not in the physical space of geography. Environmental gradients are not the same as spatial gradients. This is not to say the geographical distance is irrelevant. Basic plant processes of growth and dispersal are local phenomena and some degree of vegetation composition at a geographical location will be determined by which species are growing close by. Indeed, it might be a good idea to remove such purely spatial patterns before deriving environmental gradients, and there is a growing literature on statistical methods to do this (Borcard *et al.* 1992, Legendre and Legendre 1998, Legendre and Anderson 1999, Dale and Fortin 2002). However, causally efficacious environmental variables also show spatial covariation at many scales (Bell and Lechowicz 1994) and one would not want to inadvertently remove these causally efficacious effects in an overzealous desire to remove the spatial effects.

These points raise some important practical questions. How should we choose geographical locations in order to best quantify environmental gradients? How should we sample the vegetation at each geographical location? Such questions have been posed by community ecologists for a long time. A good review of this literature, along with recommendations, can be found in Kenkel *et al.* (1989). I can only suggest some general principles based on what we are attempting to achieve. The environmental gradients that I use in the model that will be developed in Chapter 4 represent examples of direct gradient analysis whose plant traits are community-aggregated functional traits.

Sampling

We want to choose a method of sampling that allows us to best describe the functional relationship between the plant property and the environmental variables; that is, to best measure the environmental gradient. This means that we must choose geographical locations that maximize the range of the underlying environmental variables composing the environmental gradient. Since environmental gradients are also potentially nonlinear, and if we don't know in which environmental conditions the gradient is changing most quickly, we will also want to evenly cover this range. For instance, if the environmental gradient was as shown in Figure 3.3 then we would make sure to sample soils whose pH spans the entire range from 2 to 8. If most soils in the region had pH values close to 6 then we must actively search for the more extreme soils. The purpose is not to get an unbiased estimate of the distribution of pH values (thus random sampling) but rather to map the relationship between pH and productivity. Once we know where the gradient is changing most quickly we would want to concentrate our sampling at these points. Thus, in Figure 3.3 we would want to sample more intensively in soils with pH between 4 and 5 or between 7 and 8.

However, since community-aggregated traits are (by definition) functions of the relative abundance of species, we do not want to sample within each location at such a large spatial scale that each quadrat has significant variation of the environmental gradient within it. For instance, if our quadrat was so large that it had soil pH ranging from 2 to 10 then both the measured vegetational productivity and the measured soil pH would be averaged, completely missing the underlying gradient. So, we want to maximize variation in the environmental gradient between locations and minimize such variation within each vegetation sample (for example, a single quadrat). Remember that an "environmental gradient" is not the same thing as an "environmental variable". The environmental gradient is the mathematical function of environmental variables that relates to the community-aggregated functional trait. This means that we cannot choose sizes of each quadrat that are smaller than the physical scale at which the plants are experiencing these environmental variables. For instance, even though soil nutrient supply rates can vary over a scale of a few centimeters (Bell and Lechowicz 1994), if plants integrate such variation over their entire root systems then we should also integrate at this scale. For herbaceous species this might be at a scale of 0.25 to 1 m^2 while trees might require 25 m^2 or more. My best recommendation (which should be tested and modified based on empirical evidence) is therefore to choose quadrat sizes at about the size of the average plant and to choose geographical locations (sites) that span the maximum range of variation of the environmental variables and that evenly cover this range.[10]

---

[10] Don't ask me what to do when the vegetation contains coexisting species with very different sizes, like trees and understory herbs. I don't have a good answer. Instead, I will throw the question back and hope that you are cleverer than I.

Choosing environmental variables

The reason for using environmental gradients is so that one can predict and explain the vegetational property. When dealing with plant communities as collections of things (species) then we are forced to quantify the vegetational property as compositional similarity and this leads to ordination methods producing direct environmental gradients like canonical correspondence analysis (Leps and Smilauer 2003) although the explanatory ability of such methods is weak. However, when dealing with plant communities as collections of properties of things (functional traits) then we will want to relate community-aggregated traits to environmental variables that represent selection pressures to which the functional traits are responding. Indeed, this is a key step in the model of community assembly that will be developed in this book (Chapter 5). We have a wealth of knowledge from physiological ecology, from agronomy, from forestry and from evolutionary ecology about what the major selection pressures are. These are environmental factors that directly affect plant growth, plant survival, plant reproduction and plant dispersal (Figure 3.1). So we want to ask two questions. (1) What are the environmental variables that best correlate with the changes in species composition that we want to predict and explain? (2) What are the functional traits whose selective value are most affected by changes in these environmental variables and whose community-aggregated values are most strongly correlated with some function of these environmental variables?

Placed in this context we would expect a causal hierarchy of environmental variables representing selection pressures whose importance varies over different spatial scales. At the highest spatial scale would be those climatic variables that drive biogeographic patterns (Woodward 1987, 1990a, 1990b, Woodward and Diament 1991). These would include temperature and precipitation, and therefore potential evapotranspiration, as well as the amount of solar radiation (Monteith 1977) arriving at different parts of the globe. Obviously, studies conducted at geographical scales for which temperature, precipitation or amounts of solar radiation do not vary would not have to include such variables.

At more local spatial scales we would expect to find particular non-resource environmental variables (for example, proportions of the year in which a site is flooded, concentrations of heavy metals, pH) and abiotic variables that determine resource availability and density-independent mortality; these correspond to the productivity (or stress) and disturbance gradients of Tilman (1982, 1988) and Grime (1979, 2001). Resource availability can be decomposed into the availability of mineral resources, water and open

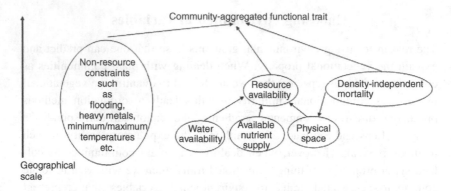

Figure 3.5. Conceptual latent environmental variables that effect functional traits at different spatial scales. These latent variables must be operationalized by appropriate indicator variables.

space[11] and each of these is then determined by individual environmental variables (Figure 3.5). It is of crucial importance that we develop methods allowing us to quantify such gradients in a way that can be (i) generalized (i.e. so that others can measure the same underlying environmental variables and apply the same function defining the gradient), (ii) that maximizes our ability to predict the traits given the environmental gradients, and (iii) that ecologists can realistically apply in the field. Soil nutrient availabilities can be estimated using ion exchange resins or soil incubation methods. Properties of soils related to water availability (field capacity, wilting point and other points along the water retention curve) can be predicted through pedotransfer functions (Bouma 1989, Pachepsky and Rawls 2004), developed by soil scientists to predict plant-relevant properties of soils from easily measured physical properties. Perhaps the most difficult environmental variables to quantify in the field are those related to the frequency and intensity of density-independent mortality (disturbance).

Statistical methods like canonical correspondence analysis can be easily modified for our purposes. The standard application of this method involves both a vegetation matrix and an environmental matrix. The vegetation matrix is usually one describing the abundance of each species in each sampling unit while the environmental matrix is one describing the values of each environmental variable in each sampling unit. However, one can simply replace

---

[11] Photosynthetically active photons, though essential for growth, are not consumable resources (Tilman 1982) because the consumption of such a photon by one plant does not, in itself, reduce the supply to other plants. Rather, competition is for space in which to be able to intercept light.

the standard vegetation matrix of species abundances by one describing the community-aggregated trait values in each sampling unit.

An alternative approach would be to more directly reflect the causal relationships between environmental variables and community-aggregated traits. This could be done using multiple regressions or statistical extensions of this using, for instance, non-parametric regression smoothers (see, for example, Shipley and Hunt 1996). Because the more causally direct environmental properties (water and nutrient availability, frequencies and intensities of density-independent mortality) are composites of individual environmental variables (soil particle size distribution, organic matter content, bedrock type, water table depth, local slope, herbivore pressures and so on) this suggests to me an approach to environmental gradients based on structural equation models where the variables inside circles are latent and would be estimated by particular observed variables that might differ from site to site (Grace and Pugesek 1997, 1998, Grace 1999, Grace *et al.* 1999, Shipley 2000a, Weiher *et al.* 2004). Whether the causal relationships depicted in Figure 3.5 are correct is another question, but one that can be answered in practice using the appropriate statistical tools.

Empirical studies of environmental gradients describing community-aggregated traits

Although physiological and evolutionary ecologists have a long tradition of describing changes in trait values along axes of environmental variables, such studies are mostly done at the intraspecific level. There are actually rather few studies that have related interspecific variation in trait values to explicitly measured environmental variables as shown in Figure 3.3, with the obvious exception of the deciduous/evergreen leaf habit that varies nonlinearly with global trends in temperature and precipitation (Kikuzawa 1991,1995, Givnish 2002). A number of studies have linked species possessing leathery leaves with a low specific leaf area to arid or semi-arid environments (Niinemets 1999, Wright *et al.* 2001) although, since such leaves are also typical of species in acidic bogs, the actual selective pressures are not simply driven by water availability (Wright *et al.* 2004). A more likely explanation is that leathery leaves with a low specific leaf area are associated with small fluxes of available (rather than total) nutrients. Before a nutrient can become available to a plant it must first be dissolved in the soil water and so arid environments do not only limit water availability but also nutrient availability. Acidic bogs, although not water limited, are strongly nutrient limited.

Wright *et al.* (2004) have shown that temperature and precipitation change the average values of a suite of leaf traits (maximum photosynthetic rate, dark respiration rate, leaf lifespan, specific leaf area, leaf nitrogen content) while these changes in average values are constrained to respect the allometric relationships between them. These very general allometric constraints have been interpreted as part of a general spectrum of traits due to a tradeoff between the ability to acquire vs. retain resources and this spectrum may, in turn, be generated by very basic constraints on the way cells and tissues can form leaves (Shipley *et al.* 2006a) . Here I will review some studies of trait–environment relationships to give you an idea of both the potential and the challenges that must be overcome.

Grime and co-workers have produced a very large data base of plant traits (Grime 2007). One of the first published examples of community-aggregated traits can be found in Grime (1974). Each species, as an adult plant, was characterized by the degree to which it expressed each of three strategies (competitor, stress-tolerator and ruderal, thus the CSR model). The degree to which a species varied along the axis from ruderal to stress-tolerator was quantified by its maximum potential relative growth rate as measured by Grime and Hunt (1975). The degree to which a species varied along the axis from ruderal to competitor was quantified by a combination of the maximum height of the leaves, the ability for lateral spread and the production of litter. More accurate rules for quantification along these two axes, involving many more traits, were subsequently developed in Hodgson *et al.* (1998). The calculated community-aggregated CSR values found in each quadrat showed clear differences in different environments. Although the environmental descriptions make it clear that the vegetation samples varied along gradients of productivity (stress) and density-independent mortality (disturbance), no actual environmental measurements were taken. For instance, sites were classified as wetland, woodland, agricultural, spoil, open habitats and wasteland.

A more recent study (Franzaring *et al.* 2007) used community-aggregated Ellenberg ecological indicator values to categorize each sample in a large data base of German vegetation. Ellenberg numbers (Ellenberg 1978) are ordinal ranks (1–9) assigned to each species that describe the level of various environmental variables at which the species reaches its maximal abundance. Significant correlations were detected between each of the six Ellenberg environmental indices (light availability (L), temperature (T), continentality (K), water availability (F), soil pH (R, soil "reaction") and nutrient availability (N)). Vegetation samples composed mostly of species typical of soils with high nutrient contents (i.e. high community-aggregated Ellenberg N

values) were also composed of species classified as either competitors or ruderals while species classified as stress-tolerators were found primarily in soils with low Ellenberg N values. Herbaceous vegetation composed primarily of species typical of wet soils (high community-aggregated Ellenberg F values) was dominated by species classified as competitors with both ruderals and stress-tolerators being associated with drier soils. Therefore, wet soils with high nutrient availability are associated with perennial plants having high potential relative growth rates, high specific leaf areas, tall stature, a rapid turnover of tissues with the ability to spread vegetatively and producing lots of litter. Drier soils with high nutrient availability are associated with annual plants having high potential relative growth rates, high specific leaf areas, short stature, a rapid turnover of tissues, a limited ability to spread vegetatively and producing less litter. Dry, nutrient-deficient soils are associated with perennial species with low potential relative growth rates, low specific leaf area, short stature, a slow turnover of tissues, limited ability to spread vegetatively, and producing less litter. Unfortunately for our purpose, Franzaring *et al.* (2007) did not give prediction equations relating various community-aggregated traits to the community-aggregated Ellenberg values and so the environmental gradients linking traits to environments are poorly defined.

Caccianiga *et al.* (2006) have applied the extended CSR classification of Hodgson *et al.* (1998) to a site in the Italian Alps that is undergoing primary succession following the retreat of a glacier. Because even the recently exposed ground had soil with appreciable nutrient levels, this sere is not as unproductive as some other classical studies of primary succession. A series of moraine ridges with known successional ages, caused by the retreat of a glacier, were sampled for vegetation (45 species) in a total of 59 quadrats (percentage cover). The authors first confirmed that the major trend in the species composition of the sites was one of successional age by ordinating the quadrats using indirect gradient methods (detrended correspondence analysis and nonmetric multidimensional scaling); in both ordinations the first axis of vegetational composition was strongly correlated with successional age. The authors then quantified the position of each of the 45 species along Grime's CSR axes using a set of 13 plant traits based on the method of Hodgson *et al.* (1998). There was a clear correlation between the successional age at which each species was most abundant and its CSR classification. The earliest sites were dominated by ruderal species, the mid-successional sites by species with intermediate ruderal – stress-tolerator values, and the oldest successional sites were dominated by stress-tolerators. Unfortunately for us, Caccianiga *et al.* (2006) did not calculate community-aggregated values of traits or even of the CSR value of each quadrat.

Diaz *et al.* (1998) studied the distribution of 19 functional traits in 13 different environmental classes, based on 100 species and 63 "non-disturbed" sites in central-western Argentina. These sites, distributed among different altitudes, varied in annual precipitation by over 800 mm and in mean annual temperature by over 11 °C and the 13 environmental classes reflected (although somewhat arbitrarily) these differences. The 19 traits were measured as ordinal variables by dividing continuous variables, when present, into classes. The authors then calculated, for each trait, the average value of each trait value in each environmental class based on presence/absence of each species; therefore these were not community-aggregated traits since rare and abundant species in an environmental class were treated as equal. To determine if the distribution of trait values differed between environmental classes they then compared the distribution in each environmental class with the distribution obtained over all classes except for the class in question. Of the 19 traits, only 4 traits showed no statistical differences between environmental classes (ability for vegetative spread, ability for carbon storage, seed shape and reproductive phenology) and 10 of these traits showed very strong discrimination. Halloy and Mark (1996) conducted a similar study of various leaf morphological traits in contrasting alpine and lowland sites in New Zealand, South America and the European Alps, and found convergence of traits in alpine vs. lowland sites at this global scale.

Lososova *et al.* (2006) studied traits that could differentiate between weed species in the Czech Republic that are typical of arable fields versus those typically associated with other human activities (roads, around buildings and so on) based on 2715 vegetation samples. After removing the effects of climate (using partial canonical correspondence analysis), they found a number of discriminating traits. Weeds of arable fields were generally annuals rather than biennials or perennials, where *R*-strategists rather than *C*-strategists (*sensu* Grime) tended to have overwintering green leaves, were insect or self pollinated rather than being wind pollinated, produced more permanent seed banks, and had limited capacity for vegetative reproduction.

Ozinga *et al.* (2004) used a huge Dutch data base (40 000 vegetation descriptions grouped into 123 "communities", involving 900 species of herbaceous seed plants) to study the relationships between traits related to seed dispersal and environmental variation in soil moisture, nutrient and light availability. Each species was coded (yes/no) as having seeds capable of being dispersed by water, wind, internally by mammals or birds, and externally by mammals (on fur). The proportion of species in each community possessing a particular seed dispersal syndrome was calculated, as well as the average value of Ellenberg indicator values for water, nutrient and light availability. Very strong relationships were found between average dispersal syndromes and these three environmental gradients. As the soil water availability increased (excluding actual aquatic or riparian communities) the proportion of species having seed adaptations to water dispersal increased ($r^2 = 0.84$). As nutrient availability increased the proportion of species with seeds adapted to wind dispersal decreased ($r^2 = 0.50$). As light availability increased the proportion of species with seeds adapted to dispersal by mammals increased while the proportion of species with seeds adapted to dispersal by birds decreased. In total, the proportions of the variance in species in a community whose seeds were dispersed by birds was 61%, by mammal fur was 56% and by mammal digestive tracts was 44% by a combination of light and nutrient availabilities.

Pywell *et al.* (2003) conducted a meta-analysis of 25 experiments, conducted in Britain, involving the restoration of grasslands on previously arable land. They quantified the abundance of each of 58 of the most common species during the first four years following sowing, corrected for treatment differences across studies ("performance"), and looked for correlations between performance at each year and 28 different functional traits. For the forbs, most traits showed significant correlations with performance, and the strength and even the direction of the correlations changed over time. Initial colonization was favored if the species was an *R*-strategist, if it has high germination percentages and especially germination in the autumn. In years two–four these traits became irrelevant and another set of traits became important: being a *C*-strategist, being able to spread vegetatively and having seed bank persistence. Stress-tolerators did uniformly poorly.

De Bello *et al.* (2005) related a series of 23 life history traits to different categories of grazing intensity in Spain. The grazing gradient was obtained using constrained canonical correspondence analysis after statistically controlling for variation in other environmental variables (mean annual temperature and precipitation, slope aspect and inclination) and the various traits were regressed on the axis related to variation in grazing intensity. This is unfortunate for us because this means that the gradient related grazing intensity to changes in species composition, not directly to changes in the traits themselves. The authors then used two different regression methods (step-wise multiple regression and non-parametric regression trees) to determine the degree to which the position of each species along the "grazing" gradient could be predicted by the plant traits. The relationships between each trait, taken singly, and the grazing gradient were significant in most cases but the predictive values (R^2) were low. The authors also did multiple regressions but did not report the predictive strength of these latter analyses.

The above studies demonstrated systematic variation between the traits of species and the types of environments in which they typically occur, but they did not produce explicit descriptions of environmental gradients, as I have defined them, because they did not produce explicit mathematical functions relating the traits to the environments. The next four studies provide sufficient information to do this.

Fonseca *et al.* (2000) sampled 46 sites in south-eastern Australia that differed in soil phosphorus concentration (P, μg g^{-1}) and mean yearly rainfall (R, mm year^{-1}). At each site they estimated the abundance of species as cover values (386 species in total) and also measured three leaf traits of each species: leaf width (*LW*, mm), specific leaf area (*SLA*, mm^2 g^{-1}) and canopy height (*CH*, m). They then calculated the average of each trait value at each site and reported the results in their Table 1. Unfortunately their reported trait means were based on simple averages in which each species was given equal weight, rather than calculating community-aggregated values by weighting by relative abundance but they say that the abundance-weighted means showed qualitatively similar behavior to unweighted means. Using these data I looked for mathematical functions relating the two environmental variables to each of the three average leaf traits using parametric multiple regression. Figure 3.6 shows the data.

The best-fitting parametric multiple regressions (i.e. the environmental gradients) for each trait are given below and the response surfaces are shown in Figure 3.7. In each case each independent variable in the regression was highly significant. This is actually our first real example of environmental gradients since we now have explicit functions linking traits (*LW*, *CH*, *SLA*) to environmental variables (*P* and *R*):

$$\log(LW) = 3.83 - 1.05 \log(P) + 0.6 \log(R)(0.3 \log(P) - 1); \quad R^2 = 0.80$$

$$\log(CH) = -14.86 + 4.45 \log(R) - 0.33 (\log(R))^2; \quad R^2 = 0.48$$

$$\log(SLA) = 11.19 + 0.11 \log(P) - 2.64 \log(R) + 0.22 (\log(R))^2; \quad R^2 = 0.59.$$

Ackerly *et al.* (2002) sampled the chaparral vegetation growing in the Jasper Ridge Biological Reserve (California), and related three community-aggregated traits (specific leaf area (*SLA*), leaf area and seed mass (*S*)) at a series of sites to the average amount of solar radiation received at the site. Figure 3.8 shows the data. Ackerly and Cornwell (2007) later report strong community-aggregated relationships with wood density and

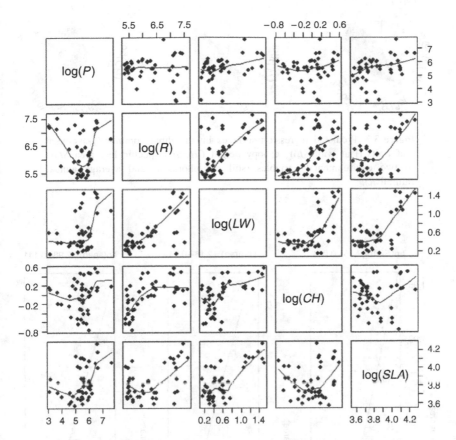

Figure 3.6. Relationships between quadrat-averaged values of soil phosphorus concentration (P, μg g^{-1}), mean yearly rainfall (R, mm year^{-1}) and three leaf traits: leaf width (LW, mm), specific leaf area (SLA, mm^2 g^{-1}) and canopy height (CH, m). Lines are from cubic-spline smoothing regressions.

plant height, and also relate the various traits to soil gravimetric water content.

Cingolani *et al.* (2007), in a study of herbaceous vegetation in Argentina, related five traits (plant height, leaf area, leaf thickness, leaf toughness and specific leaf area) to two indirect gradients obtained using detrended correspondence analysis. The first two axes were related to water availability (axis 1) and grazing intensity (axis 2). Note that these indirect gradients were constructed based on species composition, not directed based on the traits themselves, and this might be the source of some of the nonlinear relationships between the community-aggregated traits and the two compositional axes.

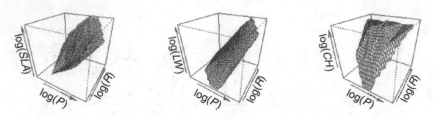

Figure 3.7. Predicted regression surfaces linking three community-averaged leaf traits (leaf width *LW*, canopy height *CH*, and specific leaf area *SLA*) to two environmental variables (soil phosphorus *P* and average annual rainfall *R*).

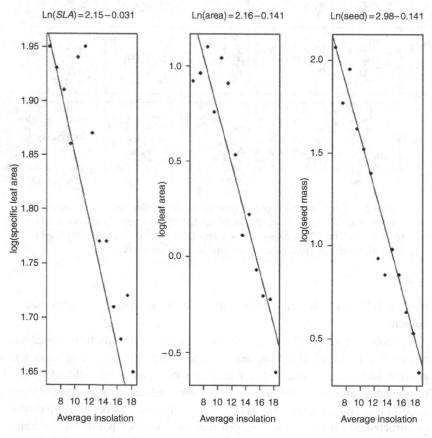

Figure 3.8. Predicted relationships between the average insolation of a site and two community-aggregated traits (specific leaf area *SLA*, seed mass *S*).

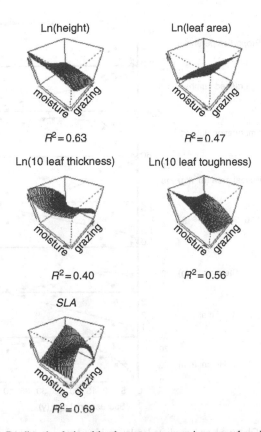

Figure 3.9. Predicted relationships between two environmental variables (moisture and grazing intensity) and five community-aggregated traits in Argentina. The proportion of the variance explained by each predicted relationship (i.e. environmental gradient) are listed below each graph (R^2).

Using floristic information from 57 samples they calculated community-aggregated values and then regressed the community-aggregated values on the two axes using multiple regression; the regression equations are reported in their Table 2 and Figure 3.9 shows the prediction equations. These regressions explained from 40% to 70% of the variance in the site scores along the two axes.

Finally, in Shipley *et al.* (2006b), Denis Vile, Éric Garnier and I related eight community-aggregated functional traits of 30 herbaceous species occurring in 10 abandoned vineyards (all within a few kilometers of each other) in southern France whose age following abandonment was known. Figure 3.10 shows the trends in the community-aggregated traits in the successional sequence. This study will be discussed further in Chapter 6.

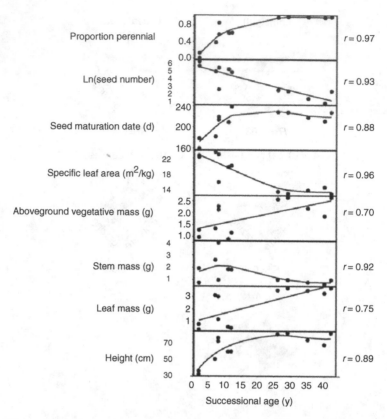

Figure 3.10. Observed and predicted relationships between eight community-aggregated traits and the successional age of each site (number of years post-abandonment). Lines show the cubic-spline regressions.

Environmental filters and assembly rules

Diamond (1975), after studying different bird communities on islands, proposed a series of "assembly rules" to account for the patterns of co-occurrence of different species on differently sized islands. Such rules took the general form of "if species X and Y are present on an island whose size is less than A, then species Z will be absent". These rules were taxonomically based rather than trait-based and the methodology was harshly criticized (Conner and Simberloff 1979). See Ribichich (2005) for a botanical example using taxonomically based assembly rules. Gilpin *et al.* (1986) attempted to derive such assembly rules for *Drosophila* species in laboratory conditions and, using a demographic approach based on Lotka–Volterra models, concluded that

"(T)*he bad news is that it is difficult to understand the structure even of laboratory communities in which one creates and controls a simple homogeneous environment, chooses species, and adds species singly, pairwise, or in higher combinations at will. After eight years of work we still have not established the relative importance of various proximate mechanisms of competition. We do not have detailed interpretations for why competitive rank shifts with food type. Our understanding of what produces the observed assembly rules is rudimentary. If these tasks are difficult in the laboratory, think how much more difficult they will be in the field, where there is an uncontrolled and heterogeneous environment, dozens or hundreds of relevant but little known species, and no opportunity for studying those species in isolation or in pairs*". This conclusion pretty much summarizes the argument that I made in Chapter 2.

Van der Valk (1981, 1988) also proposed a rather different type of an assembly rule without actually calling it that. He did this to solve a problem in the ecosystem management of prairie wetlands: how to predict the species composition that will result from changing the flooding regime. Rather than seeing wetland vegetation as being composed of things (species), leading to a demographic model, he instead concentrated on the properties of these species, specifically their germination requirements. Since water levels fluctuate greatly over time in such prairie wetlands, one obvious property that should determine species composition would be whether the seeds of a species can germinate under water or at a specified soil water matrix potential. This suggested a simple rule: starting from some initial pool of S species, remove those species whose germination characteristics prevent germination under either flooded or exposed soil, depending on the management regime.

Keddy (1992) put together Diamond's goal of deriving "assembly rules" and van der Valk's approach based on plant traits and proposed the idea of community assembly through environmental filters.[12] Think about what must happen for a plant to be found as an adult in a marsh that experiences fluctuating water levels. It must first be able to germinate on mud when water levels are low and this implies certain germination traits. This is a first environmental filter and if the plant fails to pass, it is not found in the actual community even though it is able to disperse into it. When water levels rise, its roots must survive the partially anoxic soil conditions, and this might require aerenchyma tissues or other adaptations that can transport oxygen to flooded roots. This is a second filter and if the plant fails to pass, it is not

---

[12] A similar, but more vague, notion can be traced at least to Clements in 1916 (Booth and Larson 1999).

Figure 3.11. An example of a trait-based environmental filter determining assembly rules.

found in the actual community. It must then be able to survive competition with surrounding plants, and this will depend on its size relative to the other plants (Keddy 2001). This is a third filter and if the plant fails to pass, it is not found in the actual community. This series of environmental challenges evokes the image of a series of filters that remove certain species while allowing others to pass through (Figure 3.11).

Keddy's idea of environmental filters tells us to identify the selective forces acting at a site and then match such selective forces with particular functional traits. Those species having functional traits that are adapted to the environmental conditions of the site "pass through" the filters and those species that are maladapted are excluded even though they can potentially reach the site. Note that (i) the environmental challenges that define a filter are, in reality, values of environmental variables (water depth, soil oxygen levels) and that (ii) the questions in Figure 3.11 relate values of plant traits (germination characteristics, presence/absence of aerenchyma, plant height) to these environmental variables. In other words, the filters are related to what we have already defined to be an environmental gradient. The difference is that the environmental gradient tells us which trait values to expect given a set of environmental conditions while the environmental filter tells us why we should expect them. This analogy of community assembly being determined by species passing through "filters" is evocative and is essentially the notion of natural selection applied to multi-species mixtures rather than to different genotypes of the same species. I will explore this relationship in more detail in Chapter 6. However, taken

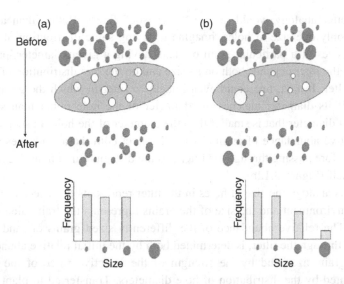

Figure 3.12. The analogy of an environmental filter. A series of plants (dark circles) belonging to different species (sizes of dark circles) form the species pool. Some trait of each species (here the size of the dark circle) interacts with a condition of the environment (the size of the open circles on the "filter") to determine if the plant will successfully "pass through" the filter. This determines the abundance of each species in the vegetation (the histograms). (a) A deterministic filter. (b) A stochastic filter.

literally, such a conceptual model is not applicable in practice. Let's see how we could modify it.

The first problem is that such a series of filters only tells us which species will occur in the community (i.e. pass through the filters) but not the relative abundance of each species. The word "filter" makes reference to the analogy of dropping a handful of sand, whose grains come in different sizes, onto a series of filters (Figure 3.12). If a grain of sand is less than the size of the holes in the filter then the grain passes. This results in the yes/no decisions of Figure 3.11. Any grain whose diameter is larger than the diameter of the holes on the filter will have no chance of passing and so would have a "relative abundance" of zero. Its "trait" (i.e. its diameter) will not be "adapted" to the selective agent of the filter (the diameter of the holes). However, any grain whose diameter is less than the diameter of the holes on the filter will be sure of passing independently of how much smaller it is than the holes in the filter. The relative abundance of those grains passing through will simply reflect their relative abundance before passing through the filter; I will call such a filter a "deterministic" filter (Figure 3.12a).

A better analogy would invoke "stochastic" filters. Rather than a filter having only one size of holes, imagine it having some distribution of sizes. The chances of a particular grain of sand, having a given diameter, passing through the filter depends both on its size and on the size distribution of holes in the filter. In fact, one could even imagine a filter in which the grain only passes if its diameter matches the diameter of the hole rather than simply having a diameter that is smaller than the diameter of the hole. In such a case the relative abundance of grain sizes will reflect both the abundances of the grains before passing through the filter and the distribution of hole sizes in the filter itself (Figure 3.12b).

In this analogy the size of holes in the filter represents the selective agents in the environment and the size of the grains represents the trait value of the plants. The relative abundance of the differently sized grains of sand, after passing through the filter, is determined both by the initial relative abundance of the grain sizes and by the strength of the selective force of the filter, represented by the distribution of hole diameters. Transferred to plant communities, the realized relative abundance of species in the community would therefore reflect the strength and types of selective forces in the environment in interaction with the different trait values of each species. Those species best adapted to the particular selective forces at work at a given point on the environmental gradient would be most abundant and those species least adapted to the selective forces at work would be least abundant.

The strength of the filter (i.e. the selective force acting at that point on the environmental gradient) is reflected in the range of probabilities possessed by the different trait values. Filter A in Figure 3.13 is a stronger filter than filter B because it more strongly discriminates between the same range of trait values.

The degree of adaptation conferred by a trait value at a given point along an environmental gradient, according to this view of stochastic filters, is measured by the probability that a plant or species possessing that value of the trait will pass through the filter. If there was only one environmental filter acting during community assembly, if there was only one trait whose values conferred a selective value with respect to this trait, and if there was a uniform distribution of species in the initial species pool with respect to this trait, then the relative abundance of species after community assembly would be equal to the degree of adaptation of the different trait values possessed by each species. (Figure 3.13).

The idea of community assembly being determined by a local species pool that is "filtered" by the interaction of functional traits with environmental selective forces has always been, for me, a seductive one. However, many

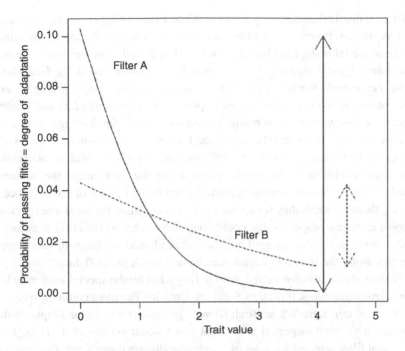

Figure 3.13. Two different stochastic filters whose discriminating ability differ. Filter A will more strongly filter species according to their trait values than will filter B.

readers will justifiably object that a seductive idea that cannot be put into practice is dangerous if it distracts us from more prosaic ideas that can be applied in practice. This is a fair criticism because there are few empirical studies that have actually identified the various environmental filters that control community assembly, quantified the strength of each relative to the appropriate functional traits, and converted these to a prediction of community structure. Before going further, I will therefore describe one study (Booth and Larson 1999) that comes close.

The Niagara Escarpment in southern Ontario (Canada) is a 450 million year old geological formation that is the edge of an ancient sea that once covered what is now the American state of Michigan. The underlying rock has been tilted and softer rock beyond it has been eroded. The result is a 725 kilometer cliff face extending from the Niagara region at the southern Canada–USA boarder north into Ontario. Niagara Falls, the largest waterfall in the world in terms of water volume, is simply the point where the Niagara River goes over this escarpment. There is very distinctive vegetation on the cliff face that is dominated by Eastern White Cedar (*Thuja occidentalis*) and this vegetation is predictable over many spatial scales. At the top and bottom (beyond the talus slope) of the cliff face, in deeper and more productive soils, are found typical forests of the region that are dominated by Red Oak (*Quercus rubra*), White Ash (*Fraxinus americana*) and Sugar Maple (*Acer saccharum*) in the south, and Aspen Poplar (*Popular tremuloides*) and White Spruce (*Picea glauca*) in the north. The ecology of this fascinating ecosystem is described in Larson (2000). Booth and Larson (1999) conducted a series of experiments to determine how the cliff environment filters out all trees except for *Thuja occidentalis*. At two locations along the escarpment the authors established five transects running from the top to the bottom of the cliff face. Along these transects they set up seed traps to determine the local tree species pool (i.e. those tree species that could potentially occur on the cliff), they took soil cores to determine the seed bank and then planted seedlings of seven tree species along the transects whose seeds could reach the cliff face.

The seed rain was dominated, not by *Thuja*, but by the species occurring in the surrounding forest typical of Southern Ontario. The most common species in the seed rain were White Birch (*Betula payrifera*) and Sugar Maple, both species with wind-dispersed seeds that don't occur on the cliff. Thus, the dispersal filter selected for a set of species producing many seeds that can be dispersed long distances by the wind. The resulting seed bank on the cliff largely reflected the relative abundance of incoming seeds. When seedlings of the different tree species were planted along the transect, all of the species could survive for the first two years but, by the third year, most had died except for *Thuja*. After harvesting the surviving seedlings of each species at the end of the third summer, the authors found that survival probabilities were related to root:shoot ratios of the seedlings. Those species dying tended to allocate more biomass to roots and *Thuja* had the lowest root:shoot ratio of all. It appears that those species that were poorly adapted to the very dry, shallow and nutrient-poor soil on the cliff face were filtered out, leaving only *Thuja*. Although the experiment was not extended into the more productive forests beyond the cliffs, it is likely that the results would be reversed.

The ability of Booth and Larson (1999) to link environments, traits and filters was due to the extreme simplicity of the system. The selective pressures acting in the extreme environment of a cliff face resulted in a single major "filter". In most real communities there will be many different correlated selective forces at work, the same trait can have different selective values with respect to the different selective forces, the traits themselves will often be correlated, and the distribution of the traits among species in the species pool will rarely be uniformly distributed. For all of these reasons, one could not equate the relative distribution of trait values to the selective value of these trait values with respect to any single environmental variable. None the less, given this analogy of stochastic environmental filters, the resulting relative abundance of trait values in the community at any instant of time should reflect (i) the combined selective value of the trait values with respect to all environmental variables, (ii) tradeoffs (due to either physical or evolutionary constraints) between the trait values, and (iii) the distribution of trait values in the species pool. I will be more precise and quantitative about this point in Chapter 5 when we link community assembly to the Breeder's Equation of quantitative population genetics. However, these points raise, to me, the most important challenge to the notion of environmental filters: how do we operationalize this idea? Even if the idea of environmental filters is appealing, how do we go from the conceptual to the practical? The test will be if we can meet the challenge posed by Keddy (2001): given (i) a species pool and (ii) an environment, can we predict the abundance of the organisms actually found in that environment?

For instance, propagules of a species must be able to disperse to a site in order to be found in the vegetation and so one could define a "dispersal" filter. Seed size would presumably be an important trait with respect to this dispersal filter for at least two reasons. First, seed mass and shape affect the chance of a seed being carried long distances by extreme wind events (Fenner and Kitajima 1999). Second, for a fixed plant mass, small seeds allow a plant to produce more of them (Shipley and Dion 1992) and this too will increase the chances of a seed successfully dispersing to a site. A second conceptual filter might determine seed survival until germination, and seed size and shape also affect the probability of a seed entering the soil seed bank and the amount of time that the seed will remain viable in the seed bank (Thompson 1987, Thompson *et al.* 1993b, 1998, Cerabolini *et al.* 2003). A third conceptual filter might determine seedling survival following germination. However, large seeds increase the probability of seedling survival when the seedling is subjected to competition for light (Foster 1986, Hewitt 1998, Fenner and Kitajima 1999, Baraloto *et al.* 2005) or when nutrients or water are limiting (Fenner and Lee 1989, Khurana and Singh 2000). In other words the same functional trait (seed size) will differentially affect the probability of the plant passing through different filters. Therefore, even if we knew how to calculate the probabilities with respect to each filter, we must then be able to combine these probabilities over all such filters.

So how does one calculate the probability (p_i) that a plant or species possessing a suite of functional traits will "pass through" each of the different stochastic filters given that traits are correlated and that environmental gradients (i.e. filters) are correlated? If we could calculate this elusive p_i for each of the S species then we would have the relative abundance of each species and therefore the structure of the community. If we could calculate p_i for each of the S species then we could also calculate any community-aggregated functional trait j as $\bar{T}_j = \sum_{i=1}^{S} p_i T_{ij}$. But we already know of a second way of determining the value of $\bar{T}_j$; we can obtain it from the environmental gradient since this is the predicted value of $\bar{T}_j$ given the values of the environmental variables. So let's turn the problem on its head. Given information of the environmental variables, we predict the community-aggregated trait value, $\bar{T}_j$, using the function describing the environmental gradient. Then, using the predicted $\bar{T}_j$ and our knowledge of the actual traits of each species (T_{ij}), we find the values of the p_i such that $\bar{T}_j = \sum_{i=1}^{S} p_j T_{ij}$. The mechanics of doing this is the subject of Chapter 4.

Divergence and convergence in trait values

If traits had no causal relationship to the probabilities of organisms being able to disperse, survive and reproduce in particular environments then the degree of resemblance of the species found in a vegetation sample with respect to such traits would be the same as that existing in the species pool except for random sampling variation. If communities are assembled by stochastic filters then those species that successfully pass through the filters will not be a random subset of the species pool and the degree to which the successful species resemble each other will be determined by how the environmental filter interacts with the traits. Filters such as those shown in Figure 3.13 will result in successful species having low values of the trait that is being selectively favored. This will result in the trait being less variable at the site than it is in the species pool (trait under-dispersion). On the other hand, classical competition theory predicts a limiting similarity with respect to traits determining competitive ability (MacArthur and Levins 1967). If this principle applies then traits determining competitive ability should be more variable at the site than in the species pool (trait over-dispersion) since species that are too similar will be competitively excluded. However, since competition between plants is a profoundly local phenomenon, and many traits could influence probabilities of survival and reproduction both through the local effects of competition with neighbors and through interactions with the physical environment, one might expect that over-dispersion, if it exists, would be more likely at small spatial scales in productive sites where competition intensity is greatest (Weiher and Keddy 1995).

To test these ideas, Weiher and Keddy (1995) measured 11 traits of a set of wetland species occurring along a gradient of soil fertility and disturbance and ordinated the species-trait matrix using principal components analysis. They then calculated the Euclidian distance between each pair of species in the space defined by the first four principal components. For each quadrat, they then calculated the average trait distance between the species that co-occurred and related this average trait distance (i.e. difference in the trait values) to the position of the quadrat along a gradient of soil phosphorus concentration. They found that quadrats occurring in infertile soils had species that were very similar in terms of the 11 traits and that the degree of difference in the trait values of the co-occurring species increased with increasing soil phosphorus levels. Using null models, Weiher *et al.* (1998) showed that those traits related to plant size were over-dispersed while the other traits were under-dispersed. Testing patterns of trait over- or under-dispersion using null models is not straightforward. Cornwell *et al.* (2006) describe a more sophisticated null model to show under-dispersion of species in a California broadleaf evergreen forest and chaparral with respect to four traits (specific leaf area, wood density, seed mass and leaf area). All of these tests suffer from the fact that they give equal weight to each species, irrespective of differences in relative abundance and therefore almost surely underestimate the degree of non-randomness that actually occurs. In any case, testing for such non-random patterns is a bit like rushing out each morning before dawn just to "test" the hypothesis that the sun will rise. If traits did not bias probabilities of dispersal, survival and reproduction of the plants that possess them, and if these biased probabilities did not differ across environments, then evolution by natural selection would be impossible.

Divergence and convergence in community trajectories during assembly

Since traits do have causal relationships with the probabilities of organisms being able to disperse, survive and reproduce in particular environments, and since trait over- and under-dispersion does occur, then we should expect systematic trends in how average trait values, and variances in trait values, change over time. This seems to mirror Clements' (1916, 1936) deterministic view of succession in which communities converge towards a common set of species with predictable dominance relationships given the same environmental conditions, irrespective of the history of community assembly. On the other hand, there is an equally long tradition in plant ecology (Gleason 1926) emphasizing the historically contingent nature of community assembly in which chance variation in the timing of arrival of different species can cause divergence in community structure even when the initial environmental conditions and the species pool is the same. Is it possible to reconcile these two seemingly contradictory views?

Yes, if we remember the difference between seeing things (species) and the properties of things (traits). The notion of stochastic filters predicts a systematic trajectory of traits, not of species. Trait combinations in a given environmental context determine the probabilities of dispersal, survival and reproduction of individuals of different species. However, since the same trait can have different selective values with respect to different environmental variables, since these traits are correlated due to tradeoffs, and since the environmental variables will also be correlated to varying degrees, this means that different trait combinations can determine similar probabilities of dispersal, survival and reproduction. Which individuals of which species that arrive at a site, and in which temporal order, will be partly determined by unpredictable events. Thus, trajectories of community development can differ even given the same initial environmental conditions and the same species pool. However, the trajectories of average trait values, being under the control of natural selection, would still show predictable changes over time. Trajectories of species abundances could show historical contingency even though these historical contingencies would be constrained to agree with the average trait values and their variances.

No one has (to my knowledge) studied this question directly, but Fukami *et al.* (2005) have done something similar. These authors seeded out vegetation in each of 15 plots (10 × 10 m) divided into three blocks of five plots each. All plots were initially similar in environmental conditions and the allocation of treatments to plots was random. Each of the five plots in each block received a different seed mix: a high diversity mixture, a low diversity mixture plus a control plot in which natural colonization occurred. The high diversity mixture consisted of five species of grasses, five species of legumes and five species of non-legume forbs. Each of the three low diversity plots (one in each block) received a different combination of two of the grass species, one of the legume species and one of the non-legume forbs. These plots were then allowed to develop naturally for nine years, including allowing natural colonization of other species over time, except that the vegetation was cut at ground level each fall. Each July (at the peak of aboveground biomass) the cover of each species in each plot was estimated using 12 1 × 1 m subplots. At the end of the nine year period the authors recorded 87 species beyond the 15 that had been sown.

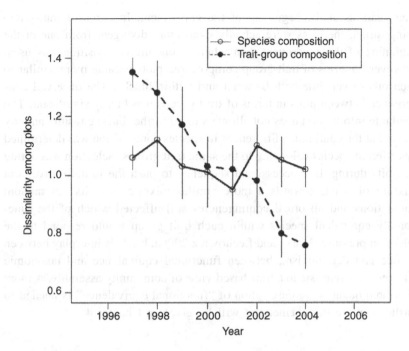

Figure 3.14. Changes in vegetation composition based on taxonomic composition and based on trait-group composition during an experimental study of
succession.

Unfortunately, the authors did not calculate community-aggregated trait
values but they did classify each species into one of 14 "trait groups" based
on functional traits, using a statistical clustering method. Because these 14
trait groups were defined by the functional groups, two plots having similar
trait-group composition will also be similar in terms of the functional traits
possessed by the species. Since information was available on both the taxonomic composition (species present and their abundances) and the trait-group
composition of each plot (trait groups present and their abundances), it was
possible to see the degree to which plots converged over time with respect to
either the species composition or the trait-group composition of the vegetation. The degree of difference between plots during each year was calculated
as a Euclidian distance using the first four principal components. Statistical
significance of changes in the degree of similarity (i.e. the Euclidian distances) was obtained using a randomization test.

What happened? Figure 3.14 summarizes the results. In terms of taxonomic composition, the different plots were just as divergent at the end of

nine years as at the beginning of the experiment. In terms of trait-group composition, the plots were actually even more divergent from one at the beginning of the experiment than when taxonomic composition was used. However, in terms of trait-group composition, plots became more similar to each other over time both between and within blocks. The observed convergence between plots in terms of trait groups was highly significant. The available information does not allow us to go further but the result can only mean that the continuing differences in species composition was determined by different species belonging to the same trait groups. Selection was acting on traits during this successional sequence to push the initially divergent mixture of plants towards a more similar mixture of traits but random fluctuations and historical contingencies still affected which of the functionally equivalent species within each trait group would respond to the selection pressure (Marks and Lechowicz 2006a, b). This interplay between chance and determinism, between functional equivalence and taxonomic difference, is intrinsic to a trait-based view of community assembly. A more exact and quantitative description of "functional equivalence" is needed to further explore this theme, and will be given in Chapter 4.

Trait variation between and within communities

The values of functional traits of species will vary both between sites (quadrats, plots) and within sites. Ackerly and Cornwell (2007) have described a method of partitioning the trait values of a given species into within-community and between-community components. However, our interest is not at the level of individual species but rather at the community level. We can do an analogous partitioning at this level. If we look at the vegetation at a single site then the species occurring at this site will differ both in terms of their abundances but also in terms of their functional traits. For example, Figure 3.15 shows the distribution of trait values in three hypothetical species at two different sites, where the size of the symbols is proportional to the abundance of the species.

The overall variation in the interspecific trait values is due to two things: (i) the variation between plants of different species that are occurring together at the same site (within-site interspecific trait variation) and (ii) the variation between plants of different species that are occurring in different sites (points along an environmental gradient). Using Whittaker's (1967, 1970) definition of α (within-site) and β (between-site) diversity, we can call this α and β interspecific trait variation. Since different species have different relative

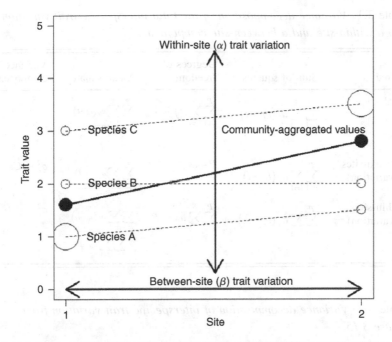

Figure 3.15. Decomposition of the variation in community-aggregated traits into between-site (β) and within-site (α) variation.

abundances, we must take this into account. The relative sizes of the circles in Figure 3.15 indicate the abundances of each species in each site. To separate these two sources of variation we can use the variance decomposition of a basic one-way ANOVA and calculate variance components. To begin, notice that we can express the interspecific trait variation of the species ($i = 1$ to S) in site 1 (S_1^2) as:

$$S_1^2 = \frac{\sum\limits_{i=1}^{s} a_{i1}(t_{i1} - \bar{t}_1)^2}{\sum\limits_{i=1}^{S} a_{i1} - 1}. \qquad \text{(Eqn. 3.1)}$$

Here a_{ij} is the abundance of species i in site j, t_{ij} is the trait value of trait t for species i in site j, and $\bar{t}_j$ is the community-aggregated trait value of trait t in site j. S_j^2 will be small if the co-occurring species of site j – especially those co-occurring species that are most abundant – have very similar trait values and will increase as these species have increasingly different trait values. Extending this across sites, we get a standard ANOVA table. Table 3.1 shows this, along

Table 3.1. *Variance decomposition of the total interspecific trait variation into a within-site and a between-site component.*

Source	Sum of squares	Degrees of freedom	Mean square	Variance component
Total	$\displaystyle\sum_{j=1}^{P}\sum_{i=1}^{S} a_{ij}\left(t_{ij}-\bar{t}\right)^2$	$\displaystyle\sum_{j=1}^{P}\sum_{i=1}^{S} a_{ij}-1$	$\dfrac{\displaystyle\sum_{j=1}^{P}\sum_{i=1}^{S} a_{ij}\left(t_{ij}-\bar{t}\right)^2}{\displaystyle\sum_{j=1}^{P}\sum_{i=1}^{S} a_{ij}-1}$	
Between-sites (β variation)	$\displaystyle\sum_{j=1}^{P}\sum_{i=1}^{S} a_{ij}\left(\bar{t}_{j}-\bar{t}\right)^2$	$P-1$	$\dfrac{\displaystyle\sum_{j=1}^{P}\sum_{i=1}^{S} a_{ij}\left(\bar{t}_{j}-\bar{t}\right)^2}{P-1}$	$\sigma_\alpha^2 + C\sigma_\beta^2$
Within-sites (α variation)	$\displaystyle\sum_{j=1}^{P}\sum_{i=1}^{S} a_{ij}\left(t_{ij}-\bar{t}_{j}\right)^2$	$\displaystyle\sum_{j=1}^{P}\sum_{i=1}^{S} a_{ij}-P$	$\dfrac{\displaystyle\sum_{j=1}^{P}\sum_{i=1}^{S} a_{ij}\left(t_{ij}-\bar{t}_{j}\right)^2}{\displaystyle\sum_{j=1}^{P}\sum_{i=1}^{S} a_{ij}-P}$	σ_α^2

Table 3.2. *Variance decomposition of interspecific trait variation from Figure 3.15.*

Source	Sum of squares	Degrees of freedom	Mean square	Variance component
Total	21.2	19	1.116	
Between-sites (β variation)	7.20	1	7.2	0.642 (45%)
Within-sites (α variation)	14.00	18	0.778	0.778 (55%)

with the variance components. In the more likely case in which the total abundance of individuals per site varies across sites one would use maximum likelihood methods to estimate the variance components.[13] Table 3.2 gives the values using the data shown in Figure 3.15 showing that, in this hypothetical example, 55% of the interspecific trait variation is due to differences between coexisting plants in the same site (α variation) while 45% is due to differences between sites (β variation).

---

[13] The easiest way to do this is by expressing the ANOVA model as a mixed model with sites defining the random component and the replicates within sites being specified by the species weighted by abundance.

Strengths and weaknesses

Let's recall the goal. A successful theory of community assembly would be able to take a list of species in a species pool, some measurable properties of these species, and a description of the environmental conditions of a site, and predict the (relative) abundance of each species that will be found at that site. Such a theory should also be able to predict how this community structure will change if the environmental conditions change, if new species are added to the species pool, or if existing species are removed. The last chapter outlined the weaknesses of a species-based approach to community assembly: the restrictive assumptions and proliferation of parameter estimates of such an approach makes it virtually impossible to confront the theoretical equations with empirical data of real plant communities. However, the species-based approach also has an important strength, namely that a theoretical infrastructure exists within which to frame empirical results.

The strengths and weaknesses of trait-based community assembly, as described in this chapter, are a mirror image of those of the species-based approach described in the last chapter. Because it is explicitly empirical, because it is based on real plant communities with all of their complications, because it is based on properties of species (i.e. traits) that are generalizable, and because a number of large trait data bases already exist, trait-based ecology is able to provide generalizable quantifications of certain community-level trends. However, nothing in this chapter will allow us to make quantitative predictions of relative abundance of each species from a species pool. Knowing that vegetation at a particular point along an environmental gradient has particular community-aggregated trait values, and perhaps also knowing the community-aggregated trait variation at that point along an environmental gradient, does not allow us to say *which* species will be present and at what relative abundance. Of course, if the environmental gradient predicts a community-aggregated specific leaf area is $100 \text{ cm}^2 \text{ g}^{-1}$ then we will likely not expect to find many individuals of a species having an *SLA* of $500 \text{ cm}^2 \text{ g}^{-1}$ at the site, but such a vague and qualitative prediction is no better than those that I criticized in Chapter 2. For all of the hand waving that I have done in this chapter, I still do not have a working theory of community assembly. The strength of the species-based approach is that there exists a formal mathematical language that can convert input assumptions into logically necessary output predictions but the trait-based approach of this chapter lacks such a formal mathematical language. The next two chapters will develop such a theoretical infrastructure.

4

Modeling trait-based environmental filters: Bayesian statistics, information theory and the Maximum Entropy Formalism

The previous chapter described a verbal model of community assembly: trait-based environmental filtering. This conceptual model views the environmental conditions of a site as a series of "filters" consisting of various selection pressures. These selection pressures reflect the probability of a given species being able to immigrate into the site and then to survive and reproduce. Such demographic probabilities vary between species because each species has a unique set of functional traits and because such functional traits bias these probabilities. The problem with this verbal model is that we don't know how to formally link traits with such probabilities. We now have to build such formal links.

With a chapter title including the words "Bayesian statistics", "information theory" and the "Maximum Entropy Formalism" you might be tempted to skip to the next one. Please resist this understandable temptation because, although this chapter is more statistical than ecological, it will develop the statistical and mathematical methods upon which the ecological theory is based. If you are a typical reader then you will be reading this book in order to learn about the links between organismal traits and ecological communities. If so then, for you, the theoretical content of the book is only a means to an end. In order to convince you that reading this chapter will be worthwhile let's be clear about what we are trying to accomplish and, equally important, what we are trying to avoid.

First, we are trying to avoid acting like a demented accountant. The patterns of community ecology are the outcome of individuals interacting with each other and with the abiotic environment. The complexity of these interactions, the idiosyncratic sensitivity of such interactions to different species and environments, and their vast number, means that any theoretical framework is doomed to collapse under its own weight if it tries to predict community patterns by accounting for each such interaction. We do not want a model that is too complex to be subject to strong empirical tests using real plant communities, as is the inevitable consequence of a demographic model of community assembly.

Second, we are trying to avoid constructing a theoretical framework based on biological assumptions that we know to be wrong. Doing so might allow us to reproduce certain community-level patterns but will not allow us to explain them and will invariably mean contradicting other known community patterns. As an example, by pretending that all individuals of all species have identical probabilities of immigration, survival and reproduction (i.e. "neutral models"[1]) one can recreate certain patterns of species abundance distributions (Bell 2000, Hubbell 2001). However, doing so not only contradicts a large body of research in population and evolutionary ecology but also leads to incorrect predictions of other well-established community-level patterns, like changes in the typical traits possessed by species in communities occurring at different points along an environmental gradient, which is a well-established empirical fact that forms the basis of functional ecology (Grime 2001).

What are we trying to accomplish then? We want a theoretical framework that produces explanatory models that can make verifiable and quantitative predictions concerning community-level patterns and that is applicable in the field to real plant communities. By "explanatory" I mean a model in which certain empirically verifiable assumptions concerning properties and processes acting at the level of individuals necessarily entail, by the rules of logic, the community-level patterns whose explanation is required. By "verifiable and quantitative" I mean a model whose predictions are numbers, not simply trends, and whose values can be compared to empirical measurements obtainable in practice. This chapter develops the tool needed for the "verifiable and quantitative" part of our model; this tool is called the "Maximum Entropy Formalism". The next chapter develops the link between our tool and natural selection and this provides the explanatory link.

Besides the lack of much ecological content, some might also be tempted to skip ahead to the next chapter because the mathematics looks difficult. From experience I know that this chapter *will* be heavy going for most ecologists but the difficulty won't lie with the mathematics; if you have had a typical undergraduate ecology education then you already know the mathematical tricks needed to follow the material. Rather than mathematical opaqueness, the difficulty will come from the unfamiliar notions attached to these mathematical tricks. This is why I will place so much emphasis on interpretation before getting down to the technical details.

---

[1] Neutral models will be briefly discussed in Chapters 5, 7 and 8.

What is a probability?

Our goal in this chapter is to understand how the Maximum Entropy Formalism (MaxEnt) works, how it can be used by community ecologists, and how to interpret what it tells us. For most ecologists the most difficult notion to grasp in understanding the Maximum Entropy Formalism is the notion of a "probability". This might sound surprising since ecologists have used probabilities for a very long time but, from the perspective of MaxEnt, they have been looking out of the wrong end of the telescope. To paraphrase an old saying, a probability is in the eye of the (Bayesian) beholder and the Maximum Entropy Formalism is a type of Bayesian statistics.

Most of you will have been taught – perhaps only implicitly – that a probability is a property of the phenomenon that you are studying but this is not true from the perspective of Bayesian statistics. In what follows, a probability is a property of you,[2] not of the event you are observing. Most classical[3] statistics courses for biologists define a probability as the limiting frequency with which a particular event occurs, relative to the frequency with which all possible events can occur, when the experiment is repeated under identical conditions an infinite number of times (von Mises 1957);[4] in fact, Kolmogoroff (1956) made this an axiom of probability. Consider what Cramér (1946, p. 154) says in the context of a six-sided die.

> *"The numbers p_r [i.e. the probability of event r] should, in fact, be regarded as physical constants of the particular die that we are using, and the question as to their numerical values cannot be answered by the axioms of probability theory, any more that the size and the weight of the die are determined by the geometrical and mechanical axioms. However, experience shows that in a well-made die the frequency of any event r in a long series of throws approaches 1/6, and accordingly we shall often assume that all the p_r are equal to 1/6 . . .".*

In frequency-based statistics, as emphasized in the above quote by Cramér, a "probability" is a property of the phenomenon, not of the observer. It is a "physical constant" of the phenomenon that is manifest as a limiting frequency and is the same no matter who is observing it. It is no more possible for two people to correctly ascribe different probabilities to the same face of a die than it is for them to correctly ascribe different surface areas to it. The surface area

---

[2] More exactly, a property of what information you have.

[3] The term "classical", though common, is misleading. Frequency-based statistics only became popular at the beginning of the twentieth century when people interested in biology, rather than physics, began doing statistical research. Earlier workers in statistical theory and its application (the Bernoulli brothers, Laplace, Poisson, Gauss, Boltzmann, Maxwell, Gibbs) viewed the notion of "probability" in a way rather closer to the interpretation given here.

[4] Second English edition of the original third German edition.

of a face on the die is a physical property that you have to measure (with some error), not something that you can deduce from some logical argument. In the same way, the probability that the face will fall up is also (using the frequency notion) a physical property that you have to measure (with some error), not something that you can deduce from some logical argument.

The notion of "probability" used in this book is a Bayesian one and is the exact opposite of how Cramér viewed a probability: the (Bayesian) probabilities p_r of our die have nothing to do with the die; rather, they refer to what we *know* about the die. A probability is a property of the observer[5] not of the phenomenon that he is observing. To explain the difference, consider a coin toss.

Imagine that I have thrown a coin 10 000 times under what I believe to be identical conditions, and find that "heads" appears with a proportion of 0.5001. If I use the frequency definition of a probability, I would say that the probability of this coin landing heads is very, very close to 0.5001. Now, if I were to throw this same coin 10 more times under the same identical conditions, what is the probability that it will land "heads" exactly four times out of 10? The answer, according to statistics texts, is to refer to the binomial distribution:

$$p(H = 4|N = 10, p = 0.5001) = \frac{10!}{4!6!}(0.5001)^4(0.4999)^6 \cong 0.2050.$$

Now, what does this answer mean? When I do the experiment (that is, I throw this same coin 10 times) then I will either observe four heads or I won't. If the answer is telling me what the actual outcome of the experiment is going to be, then it has to be either the mathematical equivalent of "you will get four heads" or else of "you won't get four heads". Since the binomial distribution has just told me that the probability that it will land heads exactly four times is 0.2050, clearly the binomial distribution is not telling me what *will* happen in the experiment! So what is it telling me?

The frequency definition of a probability is telling me the following: If I were to do my experiment of throwing the coin 10 times, and then repeat it a very large number (actually, an infinite number) of times, then 20.50 % of these repeated experiments would have exactly four heads. But this is an answer to a question that I didn't ask! I didn't want to know the proportion of times that I would observe four heads in 10 throws after repeating the experiment for the rest of eternity; I only asked about the "probability" of observing four heads out of 10 throws if I were to perform the experiment

---

[5] More specifically, since we will be dealing with an objective Bayesian definition, a probability is a property of the information available to the observer (either human or machine) about an event.

once. Therefore, the frequency interpretation of "probability" cannot be the meaning that I had in mind when I asked my question.

Maybe what I really wanted to know is the following. Given that (i) I already know that this coin lands "heads" with a proportion of 0.5001 in repeated trials, and (ii) that it will be thrown in exactly the same way 10 more times, and (iii) *that I know nothing else than points (i) and (ii) about either the coin or the throws*, what weight should I place on the claim (the hypothesis) that, when I actually perform my experiment once, I will observe four heads out of 10 versus alternative hypotheses that I will observe other proportions of heads? The conditions (i), (ii) and (iii) are called *prior* information; that is, information that I had before doing the experiment. If so, then by "probability" I mean "the weight that I should place on a particular hypothesis, given specified information". This is a Bayesian interpretation of "probability".

This Bayesian "probability" is a property of what I know about the coin and the experimental setup, not a property of how the coin actually behaves in repeated trials. Furthermore, it is still not a description of what *will* happen when I do the experiment for two reasons. First, remember that the outcome of the experiment will either be "four heads" or "not four heads". Saying that the probability of four heads in 10 throws equals 0.2050 still makes no sense if it is a description of what will occur. Second, if two people possess different types of prior information then the two Bayesian probability estimates must be different even though the eventual outcome of the 10 throws will be the same for the two people. A probability then, in this Bayesian context, is not a description of what will occur but rather is a weight ascribed to a hypothesis conditional on particular knowledge.

What do I mean by placing a "weight" on a hypothesis? Notice that, in reality, there are 11 mutually exclusive possible outcomes to my experiment $(0, 1, \ldots, 10$ heads out of 10 throws) and so there are 11 mutually exclusive hypotheses about what will happen. Imagine that I have to choose one of these 11 hypotheses before performing the experiment and that, after the experiment is finished, I will have to pay a penalty that is a function of how far my prediction (i.e. the hypothesis that I choose) differs from the outcome. This is quite close to how scientists actually function. We each make some claim about Nature based on our present knowledge and then pay a price (or gain a reward if you prefer to think positively) that increases as our predictions deviate farther from what Nature reveals. For instance, people who are spectacularly and consistently wrong are called "cranks" and we discount their authority. More positively, people who consistently get it right are ascribed wisdom.

So, imagine that we play the following game. We are asked about the outcome of some new event about which we have some prior information and then we are given a fixed amount of some resource that we value highly. We consider all mutually exclusive outcomes that are logically possible, stated in the form of hypotheses, and then we bet a proportion of our resource on each hypothesis such that all of the resource has been bet. Nature then reveals the actual outcome of the event and gives us a reward equal to the amount that we bet on each hypothesis. The game isn't finished. We are continually confronted with new events and are forced to place bets our entire life. The *real* object of the game is to get the most rewards possible (to maximize our predictive ability given incomplete prior information). We can think of the amount that we risk[6] by betting on each hypothesis as a measure of the "weight" that we place on it.

Let's define the "weight" of a hypothesis given prior information as a numerical expression of the proportion of the total resource that I am willing to bet on a given hypothesis in such a way that, over the long run, I maximize my payoff; that is, I maximize my predictive ability given what I know. If so then *how* should I allocate this resource to different hypotheses in order to maximize my payoff? Jaynes (2003, Ch. 1–2) provides the five necessary rules to do this based on the following properties:

Property I. The weights are represented by real numbers such that an infinitesimally greater weight will be represented by an infinitesimally greater number (i.e. the function mapping weights to hypotheses must be differentiable over its range).

Property II. The weights should qualitatively agree with how people reason from imperfect knowledge. This goes all the way back to Aristotle who described the rules for deductive reasoning (reasoning from perfect knowledge) versus inductive reasoning (reasoning from imperfect knowledge). You can think of deductive logic as the logic that one would apply in a system in which we knew all relevant variables and all of the logical relationships between them (i.e. when we have perfect and complete information). In such a case we would always end up with a true/false conclusion with no uncertainty (i.e. Boolean logic) and would therefore be reasoning from perfect knowledge. Inductive logic is the logic that one would apply in a system in which we only know some of the relevant variables and some of the logical relationships, but also know that there exists some variables and some relationships of which we are ignorant. In such a case we could never end up with a true/false conclusion; we would always be faced with some uncertainty and so would be reasoning from imperfect knowledge. In Figure 4.1 what we

---

[6] This thought experiment about "bets" is just an analogy and shouldn't be taken too far. We will see later that we are really talking about objective logical inferences given incomplete knowledge.

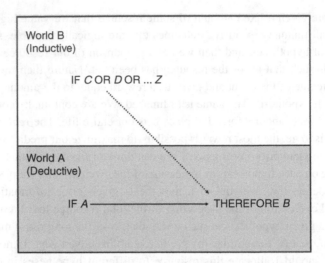

Figure 4.1. Schematic representation of the difference between inductive and deductive inference. The gray area is the area about which we have information. If this is the complete information (world A) then the inference is deductive. If this is not the complete information (world B) then the inference is inductive.

know (or prior knowledge, i.e. "if *A* then *B*") is shown in gray. We would use deductive logic if we were living in world A but inductive logic if we were living in world B.

The two basic rules of deductive reasoning (that is, reasoning from within world A in Figure 4.1 in which we *know* that *A* and *B* are the only two variables that exist) are as follows.

(i) Given: If *A* is true, then *B* is true. Given: *A* is true. Deduction: Therefore *B* is true.

(ii) Given: If *A* is true, then *B* is true. Given: *A* is false. Deduction: Therefore *B* is false.

The three rules of inductive reasoning (reasoning from world B in Figure 4.1, in which we know about *A* and *B* but our imperfect knowledge prevents us from knowing if there are other relevant variables as well) are as follows.

(i) Given: If *A* is true, then *B* is true. Given: *B* is true. Inference: therefore *A* becomes more *plausible* (i.e. we will ascribe a greater weight to the hypothesis that *A* is true once we know that *B* is true). We don't know what other relationships *B* might have with the world and so it is possible for *B* to be true even if *A* is false but, given what we do know, we increase our weight on the hypothesis that *A* is also true.

(ii) Given: If A is true, then B is true. Given: A is false. Inference: therefore B becomes less plausible (i.e. we will ascribe less weight to the hypothesis that B is true). We don't know what other relationships B might have with the world and so it is possible for B to be true even if A is false but, given what we do know, we increase our weight on the hypothesis that B is also false.

(iii) If A is true, then B becomes more plausible. B is true, therefore A becomes more plausible.

Property III. If a conclusion can be reasoned out in more that one way given the same information, then every possible way must lead to the same result. This simply states that if we logically assemble different bits of the same prior information in different orders, these logical chains cannot lead to internal contradictions.

Property IV. One must always take into account all of the available (prior) evidence that is relevant to the plausibility of the hypothesis but not invent evidence that is not available. This is simply to emphasize that we are attempting to objectively quantify information relative to hypotheses, not attempting to lie or to fool others into preferring particular hypotheses. In other words the attribution weights (plausibility) must be completely fair.

Property V. Equivalent states of knowledge must be given equivalent weights. This is again to avoid logical contradictions. Since we are trying to quantify information relative to hypotheses, two people (or machines) having exactly the same prior information cannot logically assign different weights to the same hypothesis; doing so would contradict property IV.

Given these five properties (axioms), Jaynes proves that the weight function for hypotheses, if it is to agree with the axioms of plausible reasoning given above, must have exactly the same properties as those of a probability distribution. In other words, Jaynes proves that the axioms of probabilities that you probably learnt (and forgot?) in your introductory statistics course are also the axioms for inductive logic (see Box 4.1 if you are interested). Stated slightly differently, the axioms of probability define an inductive logic.

Box 4.1. Properties of the weight function, $w(\ )$, derived from the five axioms of Jaynes

The notation $w(A|B)$ means "the weight that we will give to the hypothesis that event A will occur, given that prior information B is true".

(1) $w(AB|C) = w(B|C) \cdot w(A|BC) = w(A|C) \cdot w(B|AC)$
(2) $w(A|B) + w(\neg A|B) = 1$
(3) $w(A + B|C) = w(A|C) + w(B|C) - w(AB|C)$

Box 4.1. (Continued)

(4) $w(A_1 + \cdots + A_j|C) = \Sigma w(A_i|C)$ if $w(A_iA_j|C) = 0$ for all pairs i, j

(5) $0 \leq w(A|B) \leq 1$

(6) $0 = w(A|B)$ only if A is impossible

(7) $1 = w(A|B)$ only if A is certain

Furthermore, from (1) it follows that $w(A|BC) = w(AB|C)/w(B|C)$ if $w(B|C) \neq 0$. The notation "$w(AB|C)$" means "the weight that we give to the joint hypothesis that A and B are both true, given that we already know that C is true". The notation "$w(\neg A|B)$" means "the weight that we give to the hypothesis that A is not true, given that we already know that B is true". Property (1) is the statement of conditional probability. Property (2) is the normalization requirement of a probability density. Property (3) is the general sum rule of probabilities. Property (4) is the sum rule for mutually exclusive events. Properties (5), (6) and (7) are the requirements that a probability be bounded by 0 and 1 with these two bounds representing impossibility and certainty respectively.

I won't get into a discussion about which definition of a *probability* is "right". Let's simply acknowledge that different definitions exist,[7] that these different definitions lead to different interpretations, and that these different interpretations lead to different ways of using and manipulating probabilities. This is what I mean in this book when I talk about a "probability".

> *A probability is a numerical expression of the degree to which a given amount of information would lead an observer, obeying the rules of inductive logic, to prefer one possible outcome of an event over all others.*

Information entropy

Defining a probability in terms of the amount of information that is available is not an advance if we don't have a formal definition of "information

---

[7] Since you are having so much fun let me just mention that even Bayesians don't completely agree! The definition I have given is an "objective" Bayesian one. "Objective" means, as one of the axioms requires, that different observers having the same information about an event must ascribe the same probability; they must be logical machines. The information going into the calculation is not a subjective opinion ("what I believe") but simply a logical manipulation of axioms and so even a computer can compute such a probability. Another school of Bayesians – "subjective" Bayesians, disagree. Then there is Popper's "propensity" definition of a probability which has never been taken up (as far as I know) by statisticians.

content". This leads to a branch of mathematics called "information theory" and to notions like information "entropy" and "relative entropy". Here we go.

Entropy, for discrete events, is defined as $H = -\sum_{i=1} p_i \ln p_i$, where p_i is the probability that we assign to the occurrence of event i in a set of mutually exclusive events.[8] If I define $I_i = -\ln p_i$ then I can rewrite entropy as $H = \sum_i p_i I_i = \bar{I}$. Written in this way, it is clear that entropy is simply the mean (expected) value of "I" found in the different possible outcomes of the event. This is pretty useless information if we don't know what "I" represents!

If I tell you that I_i is the amount of new information that we gain *after* learning that the outcome (the state) of a random event is i, relative to what we knew about the event before we observed the actual value, then this would be a bit more useful, but probably not much. Equivalently, I_i can be described as the degree of uncertainty that we have concerning which of the i possible states will occur *before* observing the outcome. Here, I will try and make these words more precise but, first, a warning. Whenever you see the word "information" in the field of Information Theory, you should either add the word "new" before it (thus, *new* information gained after observing the outcome, not the information you *already* had before observing the outcome) or else replace "information" with "prior uncertainty". Contrary to the usual meaning of the word, "information" is not what you know beforehand but, rather, what new things you will have learned after the fact. Therefore, the less certain that you are about a specific result before you actually observe it – the more uncertain you are about it – then the more (new) information that you will gain once you are given the answer. To me the word "information" was a poor choice of words but we are now stuck with it. In French the information function ($-\ln(p_i)$) is called the *"function surpris"* – the surprise function – and I rather like this evocative term. I remember it as follows: *you have various levels of uncertainty (remember that we are dealing with reasoning from imperfect knowledge) before you are given the answer (the actual state that occurs). After the answer is revealed you have certainty. What you have learned, i.e. the difference between what you knew before and after, is the amount of information gained.*

Consider cold fusion. Did you know that I can create a nuclear fusion reaction at room temperature and pressure from a device that uses a single 1.5 V AAA battery? Before reading further, decide how likely this seems to

---

[8] More specifically, a series of discrete, finite and mutually exclusive events. We will see later that if the events are continuous and potentially infinite then we must use the notion of relative entropy.

you (I won't be offended). Before you actually watch me attempt this amazing feat then my claim can have two possible states (true, false). Because you do not have complete knowledge of the world, you cannot be absolutely certain about the truth or falsity of my claim, although you will have some knowledge that is relevant to it. How likely do you think it is true *before* you watch me attempt to create cold fusion? Choose a number between 0 and 1, where 0 means "it is impossible for this claim to be true and I am willing to bet everything that I value in the universe on it" and 1 means "it is certain that this claim is true and I am willing to bet everything that I value in the universe on it".

How *surprised* would you be if, after you see the result, you know that my claim is indeed false (i.e. that the number that you should have chosen was 0)? How much new information would you have gained after actually seeing me fail? That is, how far was the number that you chose from 0? Presumably, given your prior knowledge about me, about nuclear physics, and about past attempts at creating cold fusion, you were pretty sure that I would fail. If so, then you were not very surprised that I failed, your chosen number was not that far from zero, and you did not learn much that you didn't already know (you didn't gain much new information). What if you actually saw me succeed in this amazing feat without cheating? If so then the initial weight that you placed on my claim – the number that you chose before knowing the truth – would probably change greatly after seeing me succeed and you will have gained a lot of new information. You were less surprised by one outcome that the other because the less surprising outcome (that I failed to produce cold fusion) had given you less *new* information.

In order to go any further, we must now quantify the amount of new information (I) that different possible outcomes of an event (x) provides to us. Rather than simply picking a number based on some intuitive feeling, as I asked you to do just now, we want to formalize this choice of numbers. That is, we want some function that will take, as input, prior information that we have about the possible states of a variable (x) and give us, as output, a number I_i that quantifies how much new information we have gained after observing the actual state $x = x_i$; a bigger value for I_i will represent more new information gained. Equivalently, given the symmetry between how uncertain one is beforehand and how much new information is gained once the answer is revealed, we want the number I_i to quantify how *uncertain* we were beforehand, given our prior information, with a bigger value for I_i representing more initial uncertainty. We are asking, in other words, for a weight function relevant to available information and we now know that such a weight is the same as a Bayesian probability. In choosing this function, let's first establish two requirements.

(1) An outcome that, before knowing its state, was assigned a lower probability (a *prior* probability or weight) will provide more new information once we learn that it actually occurs. This is why, if you really did learn that I can create cold fusion using a small battery, you would be more surprised that if you learn that I (with no training in physics) have simply made a mistake. Formally, our first requirement is: if $p(x_1) < p(x_2)$ then $I_1 > I_2$. This is a bit counterintuitive but the key is to remember that I_i is not measuring the truth of the claim that $x = x_i$ but rather the amount of new information that would be gained *given* that $x = x_i$ is true.

(2) If two events (x and y) are independent (i.e. the outcome of one event provides no clues as to the outcome of the other event) then the information that we gain once we know the outcomes of the two events is the sum of the information that we gain by knowing each one separately. If this were not the case then one of the pieces of information must be giving us some clues as to the outcome of the other and they would not be independent. Formally, our second requirement is: $I(x, y) = I(x) + I(y)$ if x and y are independent events.

The only function involving probabilities (weights in the Bayesian sense) that has these two properties (Shannon and Weaver 1949) is: $I_i = \ln(1/p_i) = -\ln p_i$. In particular, you will notice that, if we can predict with absolute certainty that the event x_i will occur even before we observe it (i.e. $p_i = 1$), then we would gain absolutely no new information once we do observe it ($I_i = 0$). Similarly, if our prior information leads us to be almost certain that the event would not occur (say, $p_i = 10^{-100}$) then we would be astounded once we actually observe it and would have gained an enormous amount, about 230.3 natural logarithm units (i.e. $-\ln(10^{-100})$), of new information. In Information Theory one usually uses base 2 logarithms and so the amount of new information that we gain is measured in units of "bits", but one can use logarithms to any base since the results differ only by a constant[9] and in this book I will always use natural logarithms unless otherwise stated.

So, now we know that the amount of new information (I_i, the information potential) that we gain after observing that event x_i occurs is: $I_i = -\ln p_i$. Since the event x can take $i = 1, 2, \ldots, K$ different states (that is, result in $i = 1, 2, \ldots, K$ different outcomes) and since each possible state

---

[9] In particular, $\ln X = \log_e X = (\ln 2)(\log_2 X)$.

would occur (according to the knowledge that we possess before actually observing it) with probability p_i, then the *average* amount of new information that we would gain after making our observation is $\bar{I} = \sum p_i I_i = \sum p_i(-\ln p_i) = -\sum p_i \ln p_i = H$. Thus, information entropy (*H*) measures the average amount of new information that we gain once we observe the actual outcome of some process that can exist in different possible states. Equivalently, entropy measures the average amount of uncertainty that we have, given our prior information, before we observe the actual outcome of some process that can exist in different possible states.

Relative entropy

Relative entropy is a more general concept than entropy. I mentioned earlier that the definition of entropy applies to problems where the different mutually exclusive states are discrete and finite. The number of faces on a die is an example. Another example is to decide which species, out of some fixed set of possible species (i.e. a species pool), one will find. However, many problems refer to states that are not finite or that are continuous; the case of species abundance distributions, discussed in Chapter 7, is an example. In such cases the definition of entropy breaks down (the function becomes infinite) and so one applies instead the notion of relative entropy, which can be used when the alternate states are either discrete or continuous and when the total number of alternate states are either finite (and known) or infinite (or finite but unknown). Relative entropy, or "Kullback–Leibler divergence" in Information Theory (Kullback and Leibler 1951, Kullback 1959), is usually represented by the symbol "D_{KL}". Since information entropy is usually represented by "*H*", and since I want to stress the relationship of relative entropy to the entropy that you already know, I will break with tradition and represent relative entropy by "*RH*". Relative entropy is defined as: $RH = -\sum p_i \ln p_i + \sum p_i \ln q_i = -\sum p_i \ln(p_i/q_i) \geq 0$. Relative entropy is equal to zero if $p_i = q_i$, where q_i is called the "prior" probability of event *i*. We see that relative entropy is the difference between two different "types" of entropy since we can also write the equation as $RH = (-\sum p_i \ln p_i) - (-\sum p_i \ln q_i)$. If we rewrite our equation in the form of information then we get $RH = \sum p_i I_i - \sum p_i I_i' = \bar{I} - \bar{I_i'}$.

Imagine that we are given some prior information in the form of some assumptions A_0 that are encoded as probabilities q_i. We are then given some additional information in the form of extra assumptions A_1 such that our prior information now includes both A_0 and A_1 and this prior information is encoded as new probabilities p_i. The amount of new information we would gain after observing the actual outcome if we were to use both A_0 and A_1, relative to using only A_0 (and therefore ignoring A_1) is called the "relative entropy". The first term $(-\sum p_i I_i)$ is the average information that we will gain once we observe the event given the information in A_0 and A_1. We are saying: "if we have incomplete knowledge in the form of information (A_0, A_1) about the event and, based on this information, assign probabilities p_i to the different possible outcomes, then the average information that we would gain after observing the outcome of the event is $\bar{I}$".

The second term $(-\sum p_i I_i')$ is the average information that we will gain once we observe the event given only the information in the form of A_0, as quantified by the probability distribution $\mathbf{q}$. We can compare $\mathbf{q}$ to a null hypothesis. We are saying: "if we know only (A_0) but we don't know A_1, then the weight that we must assign to the hypothesis that the outcome will be i is $\mathbf{q}$. Given these null assumptions (i.e. that A_1 is irrelevant), then the expected amount of information that we would gain after observing the outcome is $\bar{I'}$". So, the relative entropy (RH) measures how much *more* information we would gain – *equivalently, how much our uncertainty is reduced*, on average, by assuming our alternative hypothesis $\mathbf{p}$ (i.e. information A_0, A_1) relative to assuming some baseline (null) probability distribution $\mathbf{q}$ (i.e. null information A_0). The relative entropy is zero only when the information content of (A_0, A_1) is equal to the information content of A_0; that is, when A_1 is completely redundant.

In using the relative entropy with respect to some process, one begins with the null hypothesis, represented by the information A_0 and encoded as the null probability distribution $\mathbf{q}$. One then adds some further information A_1 to form a new hypothesis, and encode this combined information (A_0 and A_1) in the probability distribution $\mathbf{p}$. If the null information (A_0) was very vague and uninformative then we would gain a lot of new information once we make our observation (because we knew so little before being given the answer). If the additional assumptions (A_1) were very precise then, given A_1, we would gain very little new information once we make our observation (because we knew a lot more before being given the answer). Remember: information gain does not measure the truth of a claim but rather how much new information we would gain, *assuming* that the claim is true. Since the relative entropy is the change $(RH = \bar{I} - \bar{I'})$ in the average information gain given the alternative hypothesis relative to the average information gain of the null hypothesis, then the relative entropy will be zero when $\mathbf{p} = \mathbf{q}$ and thus when $A_0 + A_1 = A_0$. In other words, the relative entropy – the change in the information gain – will be zero when the extra assumptions (A_1) included in the alternative hypothesis are irrelevant. If, besides what you already knew about me and the physics of cold fusion (i.e. A_0), you add the information that my eyes are blue (i.e. A_1), then this new information will not change your state of uncertainty before watching me perform my demonstration. Knowing the color of my eyes would not have changed your degree of uncertainty about my abilities to produce cold fusion. If the new information included in the alternative hypothesis is relevant then, assuming that it is true (which we must in calculating the entropy), it will provide less *new* information (you will learn less that you didn't already know) that does the null information once we make our observation and so the relative entropy will be greater than zero. If, besides what you already knew about me and the physics of cold fusion (i.e. A_0), you add the information that I have been supplying the electricity used by the city of Montreal from my basement, then this new information will change your state of uncertainty before watching me perform my demonstration and so this new information is relevant.

This leads to four important conclusions.

(1) The relative entropy is always greater than, or equal to, zero.
(2) If the added information of the alternative hypothesis is irrelevant; i.e. it does not change our level of uncertainty; then $\bar{I} = \bar{I}'$, $\mathbf{p} = \mathbf{q}$, and the relative entropy is zero.
(3) If the added information of the alternative hypothesis provides perfect knowledge (i.e. it allows us to perfectly predict the outcome of the event before we observe it), then there is no new information gained once we observe the result ($\bar{I} = 0$). In this case the relative entropy will be at its maximum.
(4) If the added information of the alternative hypothesis provides some, but not perfect, knowledge that allows us to predict the outcome of the event better that does the information in the null hypothesis, then the relative entropy will be positive but closer to zero than in case (3).

Eventually, we will want to apply the Maximum Entropy Formalism to relative entropy. That is, we will want to choose a probability distribution ($\mathbf{p}$) that is as uninformative as possible but that is still consistent with the facts that we know about our event. Why? Remember property IV of our rules of inductive reasoning: "One must always take into account all of the available evidence that is relevant to the plausibility of the hypothesis but not invent evidence that is not available". By choosing a probability distribution that is as uninformative as possible, but that is still consistent with what we know, we are taking into account what we know (the available evidence) but are not inventing evidence (information) for which we have no knowledge. We can do this if we choose a null distribution (thus **q**) that is as uninformative as possible (i.e. that includes the least constraining assumptions as possible) and then choose **p** to be as close as possible to **q** while still respecting the added constraints (A_1) that encode the facts we know to be true.

Continuous variables

Entropy, defined for discrete events with a known upper bound (like the faces of a die) is $H = -\sum_{i=1}^{N} p_i \ln p_i \leq \ln(N)$. One might think that, if we are dealing with continuous variables, we can simply replace the summation sign (Σ) by an integral sign ($\int$). This is wrong (doing so results in "differential" entropy). The reason comes from the difference between a probability distribution and a probability density. Since there are an infinite number of unique values for a continuous variable, the probability of observing a specific unique value is infinitely small. Because of this a continuous variable has a probability defined only over some interval. If we consider some infinitely small interval around a unique value of a continuous variable (x) and call it "dx", then the probability of observing x within this interval is (approximately) given by the area under the curve at $p(x)$ between x and $x + dx$. This is the area of an infinitely thin rectangle; i.e. $p(x)dx$ (Figure 4.2). If we replace

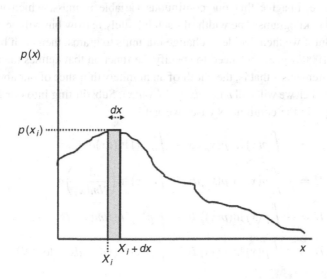

Figure 4.2. Given a continuous variable x, the probability $p(x)$ of observing a value in the interval from x to $x + dx$ is equal to the shaded area under the curve.

"p_i" in the formula for entropy in the discrete case with "$p(x)dx$" in the continuous case then we get

$$H = -\int (p(x)dx)\ln(p(x)dx) = -\int p(x)\ln(p(x)dx)dx$$

$$H = -\int p(x)(\ln(p(x)) + \ln(dx))dx$$

$$H = -\int p(x)\ln p(x)dx + \int p(x)\ln(dx)dx.$$

This looks suspiciously like our formula for relative entropy in the discrete case if we can relate "$\ln(dx)$" to "$\ln(\mathbf{q})$". In fact, they are the same thing. You can understand this intuitively if you remember the idea of a limit in differential calculus. In the continuous case we imagine constructing a histogram of the proportion of events that we observe by forming bins of our continuous variable (a histogram) and then making the bins (the range of the continuous variable that defines the width of a bar of the histogram) smaller and smaller while making more and more of them. The limit occurs when we have created an infinite number of bins, each of infinitely narrow width dx. As we approach the limit of an infinite number of bins then we have an infinitely large number (N) of bins, each of which has an infinitely narrow width (dx) which is proportional to $1/N$ (the number of bins). The width of each bin is only proportional to, but not equal to, $1/N$ because, if we change the scale of our measurement, we change the width. For instance, imagine that our continuous variable is mass, which we have measured in kilograms. The width of each infinitely narrow bin will be equal to $1/N$ kg. But if we then decide to change our units to grams, then each bin has a width of $1000/N$ g. So, we need to specify the function that defines the scale of our measurements – that is, the width of an infinitely thin slice of our abundance variable, which we will call $m(x)$: $dx = 1/Nm(x)$. Substituting into our formula for entropy in the continuous case, we get

$$H = -\int p(x)\ln(p(x))dx + \int p(x)\ln(dx)dx$$

$$H = -\int p(x)\ln(p(x))dx + \int p(x)\ln\left(\frac{1}{Nm(x)}\right)dx$$

$$H = -\int p(x)\ln(p(x))dx - \int p(x)\ln(Nm(x))dx$$

$$H = -\int p(x)\ln(p(x))dx - \int p(x)\ln(m(x))dx - \log(N)$$

$$H - \log(N) = -\int p(x)\ln(p(x))dx - \int p(x)\ln(m(x))dx.$$

(Maximally) uninformative priors

Now you can start to see the way we use information in the Bayesian context. We start with some initial, but imperfect, knowledge (A_0) and use this to obtain some initial probability estimates $(\mathbf{p_0})$. We then look at the data (i.e. get some new knowledge, A_1) and revise our initial probability estimates to get new probability estimates $(\mathbf{p_1})$ that reflect both the initial and the new knowledge $(A_0 + A_1)$. The knowledge that we had before looking is called prior knowledge and the probability estimates $(\mathbf{p_0})$ that encode this prior knowledge are called prior probabilities (or simply a "prior"). The knowledge that we have after looking $(A_0 + A_1)$ is called the posterior knowledge and the probability estimates $(\mathbf{p_1})$ that encode this posterior knowledge are called the posterior probabilities. Of course, if we then obtain even more new knowledge (A_2) then we can define a dynamic: $A_0 \rightarrow A_0 + A_1 \rightarrow A_0 + A_1 + A_2 \rightarrow \cdots \rightarrow A_0 + A_1 + A_2 + \cdots + A_n$. What was posterior knowledge at time $n - 1$ becomes prior knowledge at time n. So where do we start our learning process? What is A_0? We can start anywhere in this chain, and we will see that it is sometimes useful to begin with some substantive knowledge,[10] but often we want to start with as little prior knowledge as possible in order to explicitly introduce empirical knowledge rather than bury it implicitly into the prior. To say that we begin with no knowledge is meaningless; what does it mean to say that I know nothing?[11] More importantly, how could one quantify (via a probability distribution) the state of no knowledge?

We begin, not with the mathematically meaningless state of no knowledge, but rather with the minimal amount of knowledge that can exist while still specifying the nature of the problem. Since our problem always involves the possible states in which an event can exist, this will define our initial knowledge (A_0) and therefore our initial prior probability. We call this our *maximally uninformative* (or *ignorance*) prior. Let's begin with the simple example of a six-sided die for which we can at least distinguish each side.

Here is what we know about this phenomenon (throwing this die) without ever seeing any "answers" (i.e. empirical information) and based solely on the way the possible states are defined: there are six different and mutually exclusive states that we can recognize when we see this die. We don't know anything else. In particular, until we look at the die (and thus gain some new knowledge), we don't know if each face is distinguished by numbers

---

[10] We will do this, for instance, in order to incorporate some assumptions of neutral models; I leave this until Chapter 8.

[11] My teenaged children sometimes think that this is a perfectly meaningful claim.

(1, 2, ... , 6), by different colors or by pictures of different species of plants. There is no information that allows us to order these six states in any natural scale (i.e. no units of measurement) and so any order we choose will be arbitrary (not based on available knowledge). For instance, we don't know if one face has a much smaller surface area than another, or whether it has a magnet on it, and so on.

Given only this initial (maximally uninformative) knowledge, and our rules of inductive logic we argue as follows. First, since we know that there are exactly six mutually exclusive states, we know that $p_1 + p_2 + \cdots + p_6 = 1$ since this is one of the properties of any probability distribution. Second, what I call face "1" you might call the "red" face and a third person might call the "*Rhinanthus minor*" face but, since all orderings are arbitrary (not based on any prior information) this means that different orderings of the faces are purely arbitrary and not based on any information. Remembering property V of our rules for inductive logic ("equivalent states of knowledge must be given equivalent weights") this means that my p_1 equals your p_{red} which equals the third person's $p_{Rhinanthus\ minor}$. Since any ordering of faces is purely arbitrary (i.e. not based on empirical or deductive knowledge) our rules of inductive logic require that every side be assigned the same probability. Combining these two rules ($\sum_{i=1}^{6} p_i = 1$ and $p_i = p_j$ for all combinations of i and j), we get our maximally uninformative prior distribution: $p_i = 1/6$ for $i = 1$ to 6. Our maximally uninformative prior is a uniform distribution between 1 and 6. In general, if our problem involves a fixed number (K) of unordered discrete states then our maximally uninformative prior probability distribution ($\mathbf{q}$) is: $q_i = m(x) = 1/K$.

The argument that I have developed for our six-sided die applies for any problem in which we know the number of distinct states but have no knowledge to impose a natural ordering of these states. However, in other problems our maximally uninformative prior will not be quite so "uninformative" since we will know a bit more about the nature of the different possible states. For instance, consider states defined by different amounts of mass. Mass has a natural lower limit (zero) and a natural ordering; the notions of "less" and "more" are implicit in the notion of "mass". Since arranging our different states of mass according to this natural order is not arbitrary, there is no logical requirement that the probabilities that we assign to them be equal. This is information that we possess from the meaning of the variable "mass" and independent of any empirical information. Note that, although mass has a natural lower limit, it does not have a natural upper limit. Similarly, we cannot know how many possible unique values of mass can exist. These points prevent us from assigning a uniform prior distribution to the prior probability that our state will be observed to have a given amount of mass.[12]

Even though the ordering is not arbitrary, the actual values that we give to mass are still arbitrary. We can measure mass in pounds, kilograms, micrograms or any other unit we choose and changing the units of measurement will change the values. There is nothing in the notion of "mass" that imposes a natural unit of measure and so, based on principle V (equivalent states of knowledge must be given equivalent weights; i.e. probabilities) we must assign probabilities in our maximally uninformative prior to different states of mass independently of the scale in which it is measured. Stated another way, the probability density (since mass is a continuous variable) must be invariant to the scale of measurement.[13]

The general solution, for variables that have a physical meaning, has recently been given by Baker and Christakos (2007), itself following from Jaynes (2003) and Jeffreys (1939), and is based on the principle of dimensional invariance: the maximally uninformative prior encodes the available information about the nature of the variable in such a way that this information is the same if we change the units of measurement. Again, this is intuitively reasonable since, if all that we know is the physical nature of our variable then this information will not change if we simply change between arbitrary units of measurement. If our variable is ordered and has a natural reference point (x_0), that is, a lower limit that is based on a physical property rather than an arbitrary choice, then $m(x) = 1/(x - x_0)$. For example, many physical properties (mass, length) have a lower boundary at 0 and so $x_0 = 0$ and this leads to the well-known Jeffery's (1939) maximally uninformative prior: $p(x) = x^{-1}$. Since, in the absence of any other information besides the nature of the physical variable, we do not know where (or if) it has an upper bound, this cannot be included in the prior.[14] Temperature has a natural boundary at $-273\,°C$ (i.e. absolute zero) and so $x_0 = -273$. If the physical property changes between two physically distinct states at some value – as opposed to simply not existing below a lower boundary – then x_0 is at this value. An example would be the variable "change of length", which changes from "elongation" to "contraction" at 0.

We are almost ready to move on to the most important concept for our purposes in this book, the Maximum Entropy Formalism. Before we do, I

---

[12] In fact, since mass is a continuous variable and there is no upper limit, it is impossible to assign a non-zero uniform probability since there will be an infinite number of them.

[13] One generally also requires the variable be invariant to changes in location as well.

[14] Some readers might notice that $x^{-1} = e^{-\ln x}$, which is a uniform distribution on a logarithmic scale, and a logarithmic transformation is invariant to changes in scale.

want to head off a potential source of confusion. In the special case of unordered, finite, discrete states (like the faces on a die or the different species in a fixed species pool) for which the maximally uninformative prior is a uniform distribution, entropy and relative entropy are the same thing, except for a scale constant. Imagine that, in the absence of any empirical information and based only on the definition of the states in a problem, we know that there are K mutually exclusive discrete unordered states; in the case of a six-sided die this would correspond to $K = 6$. We know that the probabilities that encode this information and nothing else – the maximally uninformative prior – is a uniform distribution with $q_i = 1/K$. Now,

$$RH = -\sum p_i \ln p_i - \left(-\sum p_i \ln q_i\right) = -\sum p_i \ln \frac{p_1}{q_i}$$

$$RH = -\sum p_i \ln p_i - \left(-\sum p_i \ln\left(\frac{1}{K}\right)\right)$$

$$RH = -\sum p_i \ln p_i - \left(\ln(K)\sum p_i\right)$$

$$RH = -\sum p_i \ln p_i - \ln(K)$$

$$RH + \ln(K) = H.$$

The fact that the relative entropy (RH) is the same as the entropy (H) plus a constant ($\ln (K)$) when the maximally uninformative prior distribution is a uniform distribution means that the probabilities (p_i) that maximize the entropy are the same as those that maximize the relative entropy.

The Maximum (relative) Entropy Formalism

You are now ready to learn the mechanics of the Maximum Entropy Formalism. The name is an historical consequence of the fact that the method (i.e. "formalism") was first derived for the special case of a uniform maximally uninformative prior, for which maximizing entropy is the same thing as maximizing relative entropy. We should more properly call it the Maximum *Relative* Entropy Formalism (and some people do) but I will follow the general consensus and call it the Maximum Entropy Formalism (or MaxEnt) in order to avoid adding even more confusion. Before entering into the details, let's first be clear about what we will be trying to achieve.

There are two parts to the Maximum Entropy Formalism: what we know and what we don't know. What we know about the phenomenon is described by (i) certain *constraints* and (ii) by the nature of the states (i.e. the maximally uninformative prior).[15] Everything that we don't know about the phenomenon is encoded by the remaining uncertainty. You already know that uncertainty is quantified by the entropy (or the relative entropy) and the greater our uncertainty, the larger the entropy (or relative entropy). Now recall Jaynes' property IV ("one must always take into account all of the available evidence that is relevant to the plausibility of the hypothesis but not invent evidence that is not available"). The first part, "taking into account all of the available evidence", means choosing probability values (the p_i values) that agree with the constraints and with the prior distribution. The second part, the requirement not to "invent evidence that is not available", means choosing probability values (the p_i values) that, while agreeing with these constraints and with the prior, will otherwise represent maximum uncertainty. Since a probability describes our state of knowledge, then choosing probabilities that are less than maximally uncertain given what we know (the constraints) means adding extra information that we don't really possess; i.e. we would be inventing evidence that is not available. We would, in effect, be lying. Remembering that the entropy is maximal when the uncertainty is maximal (entropy is a measure of uncertainty), this means that we must choose values for our probabilities (the p_i values) that maximize the entropy (the average amount of uncertainty) while agreeing with what we do know (the constraint values and the prior). We don't choose the distribution having maximal entropy consistent with the constraints because this distribution has higher predictive value than the others but rather because we don't know any better. Think of it this way: imagine that I tell you that I know nothing about a particular die except that it has six sides. After telling you this, I proceed to bet you everything that I own on the hypothesis that the next throw will be a "6". I would be claiming ignorance but acting as if I had inside knowledge. You would likely conclude that I am either acting irrationally or else that I am lying about the amount of information that I have about the die. Since both of these possibilities contradict the principles of inductive logic they would both be excluded using the Maximum Entropy Formalism.

From the perspective of Information Theory, this is really all that you have to know in order to understand the Maximum Entropy Formalism. However, it is also possible, and perhaps more comforting for some readers, to link entropy maximization to a more physical process, although not every problem of inference that is applicable to the Maximum Entropy Formalism can be formulated in just this way. I also hasten to add that the formulation that

[15] Or, more generally, by the prior distribution since this does not have to be a maximally uninformative one.

follows is not a *derivation* of the Maximum Entropy Formalism; MaxEnt applies even when the substantive assumptions of this formulation don't hold because, in the end, MaxEnt is a principle of logical inference from imperfect knowledge not a description of any physical process. I will use the example of a six-sided die and then expand this to more general cases.

A physical model of entropy maximization

Imagine that some physical process results in a set of N objects that are distributed into K different groups. An example might be the N throws of a die resulting in a distribution of faces, the N throws of a coin resulting in a distribution of heads and tails, the growth of N individual plants resulting in a distribution of individuals per species or the capture of N resource units resulting in a distribution of biomass per species. Now we, as observers, can see the resulting distribution (the *macrostate*) but we cannot see the dynamics (i.e. the sequence of throws, the sequence of individuals growing) of how this distribution arises (the *microstate*). Many different sequences (microstates) can result in the same final distribution (macrostate) but we, as observers, cannot distinguish between different microstates that give the same macrostate.

So, at the end of N repetitions (throws of the die, captures of resource units . . .), we will observe a particular partition of the N objects into each state. We would observe n_1 objects displaying state 1, n_2 objects displaying state 2, and so on. We observe, in other words, the macrostate distribution: $\mathbf{n} = \{n_1, n_2, \ldots, n_K\}$. By dividing by N we could equivalently describe this partition in terms of proportions: $\mathbf{p} = \{p_1, p_2, \ldots, p_K\}$. As a simple example, consider a coin ($K = 2$ states) that has been thrown four times ($N = 4$ repetitions). The macrostate $\mathbf{n} = \{4\text{H}, 0\text{T}\}$ corresponding to four heads and no tails can only occur in one way (i.e. it corresponds to only one microstate) because, if we order each throw in time, then the only way to get this macrostate is: $\{\text{H}, \text{H}, \text{H}, \text{H}\}$. The macrostate $\mathbf{n} = \{2\text{H}, 2\text{T}\}$ can occur in six different ways: $\{\text{H}, \text{H}, \text{T}, \text{T}\}$, $\{\text{H}, \text{T}, \text{H}, \text{T}\}$, $\{\text{H}, \text{T}, \text{T}, \text{H}\}$, $\{\text{T}, \text{H}, \text{H}, \text{T}\}$, $\{\text{T}, \text{H}, \text{T}, \text{H}\}$ and $\{\text{T}, \text{T}, \text{H}, \text{H}\}$. Since we can only see the resulting macrostate but not the temporal dynamics of the microstate, these six different microstates are indistinguishable to us. If each microstate was equally likely then we would see the macrostate $\{2\text{H}, 2\text{T}\}$ six times as often as we see the macrostate $\{4\text{H}, 0\text{T}\}$. Table 4.1 lists each possible macrostate and the number of indistinguishable microstates that give rise to each macrostate.

The total number (W) of ways (i.e. the number of indistinguishable microstates) in which different dynamics involving N steps will lead to the

Table 4.1. *The relationship between macrostates and microstates when throwing a coin four times.*

Distinguishable macrostate	Heads	Tails	Indistinguishable microstates giving rise to the same macrostate
1	4	0	1
2	3	1	4
3	2	2	6
4	1	3	4
5	0	4	1

same macrostate $\{n_1, n_2, \ldots, n_S\}$ is given by:

$$W = \frac{N!}{n_1! n_2! \ldots n_K!}$$

or, equivalently, by: $\ln(W) = \ln(N!) - \sum_{i=1}^{K} \ln(n_i!)$. For large numbers, Stirling's approximation states that $\ln(n!) \approx n \ln(n) - n$. Using this, and the fact that $N = \Sigma n_i$, gives:

$$\ln(W) \approx N \ln(N) - \sum_{i=1}^{K} n_i \ln(n_i)$$

$$\ln(W) \approx N \ln(N) - \sum_{i=1}^{K} \left(\frac{Nn_i}{N}\right) \ln\left(\frac{Nn_i}{N}\right)$$

$$\ln(W) \approx N \ln(N) - N \sum_{i=1}^{K} p_i (\ln N + \ln(p_i))$$

$$\ln(W) \approx N \ln(N) - N \ln(N) \sum_{i=1}^{K} p_i - N \sum_{i=1}^{K} p_i \ln(p_i)$$

$$\frac{\ln(W)}{N} \approx - \sum_{i=1}^{K} p_i \ln(p_i).$$

Notice that the formula for $\ln(W)$, the logarithm of the number of indistinguishable microstates leading to the same macrostate, is proportional to the information entropy of the distribution of observations into the possible macrostates. Using the formula $W = e^{-N \sum_{i=1}^{K} p_i \ln p_i}$ we can explore how many different dynamic processes of allocation lead to same macroscopic result. Figure 4.3 plots the results for three different scenarios of throws of a coin

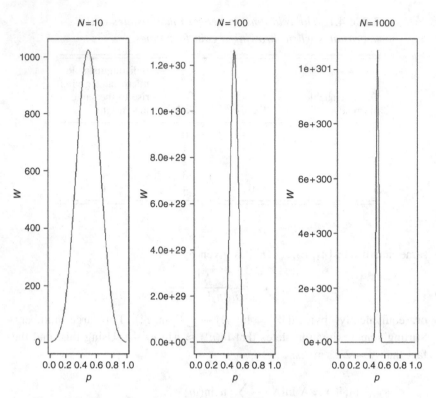

Figure 4.3. The number of ways (W) in which different microstates (the ordering of coin tosses in time) will give the same macrostate (the proportion of heads, p). Left to right shows the distribution of W when we throw 10, 100 or 1000 coins; note the huge change in scale for the abscissa from left to right.

($K = 2$) when the total number (N) of throws is either 10, 100 or 1000. One could equivalently consider cases of N throws of a die, or N individuals distributed among species, or N resource units captured and converted into biomass among species.

Figure 4.3 shows an important result; note that the abscissa goes from a maximum of 1000 (left) to 10^{301} (right). Since we are dealing with a coin having only two states (heads/tails) the macrostate distribution (the probability of a coin showing a head or a tail) can be expressed as a single number: $\mathbf{p} = \{p, 1-p\}$ and the closer p is to 0.5 the more uniform is the macrostate distribution; that is, the closer $\mathbf{p}$ is to $\{0.5, 0.5\}$. In Figure 4.3 see that, as the number (N) of sequences in the dynamics of allocation increases (going from left to right), the number of different microstates (W) increases much more rapidly in those macrostates that are most evenly distributed; i.e. those macrostates closest to $p = 0.5$. At $N = 1000$ we see that almost all different dynamic allocations would result in a macrostate of approximately equal proportions of heads and tails. In other words, as the number of different "steps" in a dynamic process increases the outcome of the process (the macrostates) become more and more concentrated in those allocations that are most uniform. If the physical properties of the process (the way the coin is tossed, the way different species can capture resources) do *not* favor some dynamics (microstates) more than others (if we have a fair coin) then the resulting macrostate distribution will be maximally uniform. If the physical properties of the system *do* differentially select some dynamics (microstates) more than others (if we are using a biased coin), then, for large N, the resulting distribution will still be highly concentrated but this concentration will occur at the macrostate that can be arrived at in the most different ways while still respecting the selection process resulting from the physical properties of the system.

This means that when a macrostate (a distribution of throws of a coin, or a distribution of biomass to different species) results from the sequential allocation of many things during a physical process then the probability of obtaining different macrostates of this process can be obtained by maximizing W subject to the physical constraints that are acting in the process. Since $\ln(W)$ is a one-to-one[16] transformation of W, one can equivalently maximize either W or $\ln(W)$. However, $\ln(W)$ is simply our formula for information entropy, multiplied by a constant (N). This means that maximizing entropy subject to constraints is the same as maximizing the number of microstates of a physical process leading to the same macrostate.

Some readers might recognize the above argument as an example of Statistical Mechanics. In fact, all of the results of equilibrium Statistical Mechanics and thermodynamics can be obtained quite directly from the Maximum Entropy Formalism (Jaynes 1957a, b, 2003) without the ancillary assumptions of dynamic ergodicity and so on. Recent work has also extended it to non-equilibrium thermodynamics as well (Dewar 2003, 2004, 2005). If you think of a collection of molecules in a gas as a "community" of molecules distributed into different energy "species", then classical statistical mechanics is a model of the structure of this gaseous community. From this perspective, and not taking the analogy too far, the approach to community structure and dynamics taken in this book is a "statistical mechanics" of ecological communities.

[16] This means that each unique value of W results in only one value for $\ln(W)$. Therefore if we know the maximum of $\ln(W)$ then we also know the maximum of W.

So, how do we go about deriving a statistical mechanics of ecological communities; that is maximizing relative entropy subject to constraints? First, let's (finally) write down our MaxEnt model of trait-based environmental filtering, and then break it down into its component steps.

The MaxEnt model of trait-based environmental filtering

Given the chosen prior distribution (**q**), the predicted relative abundances (**p**) of each of the S species in the species pool are those that maximize the relative entropy (Equation 4.1):

$$\max\left(-\sum_{i=1}^{S} p_i \ln\left(\frac{p_i}{q_i}\right)\right) \qquad \text{(Eqn. 4.1)}$$

while simultaneously agreeing with the following K community-aggregated trait values ($\overline{T}_j$) plus the normalization constraint:

$$\sum_{i=1}^{S} p_i = 1$$

$$\sum_{i=1}^{S} p_i t_{i1} = \overline{T}_1$$

$$\cdots$$

$$\sum_{i=1}^{S} p_i t_{iK} = \overline{T}_K.$$

Step 1: Quantifying what we know

In the case of maximum ignorance (which we encode using our maximally uninformative prior) we don't know any attribute of our different possible states. For our die, this means that we don't know anything about each side. For species in a species pool this means that we don't know anything about the properties of these species. The first step in moving beyond maximal ignorance is therefore to describe the attributes of the different states for which we have knowledge. In the case of a die with different sides the most basic attribute of a side is that it exists and so we can ascribe a value of $E_i = 1$ for each side that exists (i.e. the first six sides) and a value of $E_i = 0$ for all sides that don't exist (i.e. all other sides). We can be sure that, when we throw our die, it will always land on a side that exists and so the average value of our first attribute (the existence of a side) will be 1.0. Translating this into an equation (a constraint equation) gives $\sum_{i=1}^{\infty} p_i E_i = 1$, thus $1p_1 + 1p_2 + 1p_3 + 1p_4 + 1p_5 + 1p_6 + 0p_j = \sum_{i=1}^{6} p_i = 1$, where j refers to all sides other than 1 to 6.

Now imagine that we also know that each side of our die has a different number of black spots (1 to 6). The number of spots (S_i) on side i is another attribute of each unique state (a "variable", that is, an attribute that *varies* over the different states). If we have empirical knowledge that, when the die is thrown many times, the average value of S is 4.2 then we can translate this information into a second constraint equation: $\sum_{i=1}^{6} p_i S_i = 4.2$. If we had empirical information about average values of other attributes of each side (perhaps its surface area, or the size of each hole into which the spot was painted, or its degree of curvature if the sides are not flat) then we would translate each of these other bits of empirical information into constraint equations; an example from a real die will be given later. Here is the information that we have about our die, translated into our constraint equations:

$$\sum_{i=1}^{6} p_i = 1$$

$$\sum_{i=1}^{6} p_i S_i = 4.2.$$

Now, we have to assign values to these probabilities (the six p_i values) such that they agree with what we know, that is, they agree with our two constraint equations. We must also assign values to these probabilities such that they do not encode any other information beyond this; doing so would mean, remembering that a Bayesian probability is simply a quantification of information content, that we would be inventing information that we don't actually possess; we would, in effect, be lying. Stated equivalently, we must assign values to the probabilities so that they are otherwise maximally uninformative (uncertain). We do this by choosing values that will maximize the relative entropy while insuring that the values agree with the two constraint equations (what we do know) and there is always only one probability distribution that maximizes the relative entropy (Shannon and Weaver 1949, Jaynes 2003). Here is how we do it.[17]

The goal is to maximize the relative entropy subject to the constraints that encode what we know about the phenomenon. This is done using the method of

---

[17] Later, I will give you a numerical algorithm and some R code, called the Improved Iterative Scaling algorithm, that does this without the calculus.

Lagrange multipliers. We first specify our objective function (Q) which consists of the relative entropy (RH) plus each of the j constraint equations multiplied by a constant, called a Lagrange Multiplier (λ_j), for which we will have to solve: $Q = RH + \lambda_0(1 - \sum p_i) + \lambda_1(\overline{X}_1 - \sum p_i X_{i1}) + \cdots + \lambda_j(\overline{X}_j - \sum p_i X_{ij})$. In this equation the X_{ij} are the i observed values of trait j (the number of dots on the die in our example) and the $\overline{X}_i$ are the observed mean values of these traits (the value 4.2 in our example). In our die example, for which the maximally uninformative prior is a uniform distribution with each $q_i = 1/6$, this would be:

$$Q = -\sum_{i=1}^{6} p_i \ln p_i + \sum p_i \ln q_i + \lambda_0\left(1 - \sum_{i=1}^{6} p_i\right) + \lambda_1\left(4.2 - \sum_{i=1}^{6} p_i i\right)$$

$$Q = -\sum_{i=1}^{6} p_i \ln p_i - \ln(6)\sum_{i=1}^{6} p_i + \lambda_0\left(1 - \sum_{i=1}^{6} p_i\right) + \lambda_1\left(4.2 - \sum_{i=1}^{6} p_i i\right)$$

$$Q = -\sum_{i=1}^{6} p_i \ln p_i - \ln(6) + \lambda_0\left(1 - \sum_{i=1}^{6} p_i\right) + \lambda_1\left(4.2 - \sum_{i=1}^{6} p_i i\right).$$

In order to find the values for p_i that maximize the relative entropy,[18] conditional on our two constraints (the part inside the brackets), we must obtain the partial derivatives for Q (this gives us the partial slopes of Q as a function of p_i) and then solve for the values of p_i when these partial derivatives are zero (this gives us the maximum of Q). So, in the case of our die example, we have six equations of the form:

$$\frac{\partial Q}{\partial p_i} = -(\ln p_i + 1) - \lambda_0 - i\lambda_1 = 0.$$

In the more general case we will have i equations, each involving the values of the K measured attributes of each of the states, of the form:

$$\frac{\partial Q}{\partial p_i} = -(\ln p_i + 1) + \ln q_i - \lambda_0 - \sum_{j=1}^{K} X_{ij}\lambda_j = 0.$$

Solving these i simultaneous nonlinear partial differential equations, each having $K+1$ parameters (the Lagrange multipliers, λ) yields the posterior probabilities (p_i) that are both consistent with what we know (the constraints) and that are otherwise maximally uninformative. These are the probabilities that we must assign to each state if we are to follow the principles of

---

[18] Since the maximally uninformative prior is a uniform distribution in this example, the relative entropy equals the entropy plus a constant ($-\ln(6)$).

inductive logic. The general solution to this system of equations is a generalized exponential distribution:

$$p_i = \frac{q_i e^{-\sum\limits_{j=1}^{K} X_{ij}\lambda_j}}{\sum\limits_{i} q_i e^{-\sum\limits_{j=1}^{K} X_{ij}\lambda_j}}.$$

This distribution is also known as the Gibbs distribution[19] in statistical mechanics. Note that when we don't have information on any attribute (i.e. we know nothing beyond what is already encoded in the maximally uninformative prior), this reduces to $p_i = q_i/\sum_i q_i = q_i$. Box 4.2 gives more details of how to derive this solution, if you are interested. We can therefore solve (in principle) these equations to get the values of the K Lagrangian multipliers λ_j. In practice, when there are more that a few constraints, we will use a numerical algorithm to estimate the Lagrangian multipliers. This algorithm, the Improved (Generalized) Iterative Scaling algorithm, is described in Box 4.3. At this point, let's leave theory for a bit and see how one would apply the MaxEnt method to actual data. We are still not ready to move on to ecological communities, but we can apply what we have just learnt to a phenomenon that is much simpler and also easier to understand. We will then see how this relates to plant communities.

Box 4.2. Maximizing relative entropy subject to constraints:

$$\frac{\partial Q}{\partial p_i} = -(\ln p_i + 1) + \ln q_i - \lambda_0 - \sum_{j=1}^{K} X_{ij}\lambda_j = 0$$

$$\ln p_i = -1 + \ln q_i - \lambda_0 - \sum_{j=1}^{K} X_{ij}\lambda_j$$

$$p_i = q_i e^{-1-\lambda_0} e^{-\sum_{j=1}^{K} X_{ij}\lambda_j}.$$

---

[19] Josiah Willard Gibbs (1839–1903) was an American who worked, among other things, on the mathematical foundations of statistical mechanics. He was the first to show how the classic equations of equilibrium thermodynamics could be derived by maximizing the function $-\sum_i p_i \ln p_i$ (now called information entropy) subject to macroscopic constraints. He was criticized at the time because this equation seems *ad hoc*.

Box 4.2. (Continued)

Normalizing the values so that $\Sigma_i p_i = 1$:

$$p_i = \frac{q_i e^{-1-\lambda_0} e^{-\sum_{j=1}^{K} X_{ij}\lambda_j}}{\sum_i p_i} = \frac{q_i e^{-\sum_{j=1}^{K} X_{ij}\lambda_j}}{\sum_i q_i e^{-\sum_{j=1}^{K} X_{ij}\lambda_j}} = \frac{q_i e^{-\sum_{j=1}^{K} X_{ij}\lambda_j}}{Z}, \text{ where Z is called the}$$

"partition function".

Now, $\sum_i p_i = 1 = \sum_i e^{-1-\lambda_0} q_i e^{-\sum_{j=1}^{K} X_{ij}\lambda_j} = \frac{1}{e^{1+\lambda_0}} \sum_i q_i e^{-\sum_{j=1}^{K} X_{ij}\lambda_j}$. From this,

we get:

$$e^{1+\lambda_0} = \sum_i q_i e^{-\sum_{j=1}^{K} X_{ij}\lambda_j}$$

$$1 + \lambda_0 = \ln\left(\sum_i q_i e^{-\sum_{j=1}^{K} X_{ij}\lambda_j}\right)$$

$$\lambda_0 = \ln\left(\sum_i q_i e^{-\sum_{j=1}^{K} X_{ij}\lambda_j}\right) - 1.$$

We now take the partial derivatives of λ_0 with respect to the other j Lagrangian multipliers (λ_j):

$$\frac{\partial \lambda_0}{\partial \lambda_j} = \frac{1}{\sum_i q_i e^{-\sum_{j=1}^{K} X_{ij}\lambda_j}} \left[-\sum_i X_{ij} q_i e^{-\sum_{j=1}^{K} X_{ij}\lambda_j} \right].$$

But we already know that: $p_i = \frac{q_i e^{-\sum_{j=1}^{K} X_{ij}\lambda_j}}{\sum_i q_i e^{-\sum_{j=1}^{K} X_{ij}\lambda_j}}$. Therefore:

$\partial \lambda_0 / \partial \lambda_j = -\sum_i X_{ij} p_i = -\overline{X}_j$. From this, we get j nonlinear equations involving j values of λ_j:

$$\frac{\partial \lambda_0}{\partial \lambda_j} = \frac{1}{\sum_i q_i e^{-\sum_{j=1}^{K} X_{ij}\lambda_j}} \left[-\sum_i X_{ij} q_i e^{-\sum_{j=1}^{K} X_{ij}\lambda_j} \right] = -\overline{X}_j.$$

The curious experiment of Johann Rudolf Wolf

Rudolf Wolf (1816–1893) was a Swiss astronomer and mathematician at the University of Zurich. His most famous scientific contribution was to the study of sunspots but, for our purpose, his most useful contribution was the results from a curious little experiment. He must have been a very patient man (or perhaps very lonely) because he conducted an experiment in which he threw a pair of die (one red and one white) 20 000 times and recorded the number of times each of the six sides landed face-up. These data are reproduced in Fougere (1988) who analyzed them, following from the original analysis of Jaynes (1983). Table 4.2 shows the results for the white die. To get a hint of how this experiment might relate to ecological communities you can consider each side as analogous to a species in a local species pool, the die as analogous to the total species pool, and the proportions as analogous to the relative abundance of each species at a site.

Can we construct a mathematical model, based on physical traits of the sides of the die that predicts these observed proportions? It would be incredibly difficult to do this by trying to derive dynamic equations that account for all of the physical forces acting on a die. In fact, it would be impossible in practice, since such equations would be very sensitive to even tiny changes in the initial conditions. Since such initial conditions will be different for each of the 20 000 throws we would expect this part of the dynamics to be unpredictable (i.e. random). Can we instead construct our mathematical model without knowing the detailed dynamics of this die as it left the hand of Dr. Wolf and fell to the table? We start by first assuming that there are no physical differences between the sides of the die (Figure 4.4a). In other words, we are assuming that we can distinguish between the sides (otherwise there would be only 1 macrostate) but that there are no properties of the sides that bias the dynamics of a throw in favor of any particular side. If this is true then the only physical constraint that we can put on the system is that the die had six sides; thus $1 - \sum_{i=1}^{6} p_i = 0$. This, of course, is the normalization constraint associated with the λ_0 Lagrange multiplier. If this is

Table 4.2. *The frequency, and the observed proportion*
(o_i) with which each face landed up in an experiment in
which a single die was thrown 20 000 times.

Side	Frequency	Proportion (o_i)
1	3246	0.162 30
2	3449	0.172 45
3	2897	0.144 85
4	2841	0.142 05
5	3635	0.181 75
6	3932	0.196 60

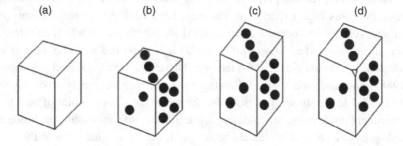

(a)   (b)   (c)   (d)

Figure 4.4. A series of possible dice having physical properties that could affect
the probabilities of different faces. (a) A perfectly fair die; (b) the result of drilling
dots into the first die; (c) the result of increasing the 3–4 axis of the second die;
(d) the result of removing a tiny chip from the 2–3–6 face of the third die.

the only physical feature of the die that determines the outcome of a given
throw, then the maximum entropy distribution given this single constraint
should agree with the observed outcomes. The maximum entropy distribution, given only this one constraint, is simply the maximally uninformative
prior which is a uniform distribution, $p_i = 1/6$, for all six sides. We can test
this using a chi-squared distribution with $6 - 1 = 5$ degrees of freedom; the
degrees of freedom equal the number of states minus the number of constraints. So, using the maximum likelihood chi-square statistic[20] we find:

$$G^2 = 2 \cdot 20\,000 \sum_{i=1}^{6} o_i \ln\left(\frac{o_i}{q_i}\right) = 269.97.$$

[20] In using this equation I am assuming that each throw of the die was independent, thus
allowing me to multiply by 20 000. This is a reasonable assumption for a die but not
necessarily for other types of allocations of observations to states.

The maximum entropy probabilities, given this constraint, are clearly different from the proportions actually observed since the null probability of such a large chi-square value, given five degrees of freedom, is much lower than 10^{-17}. Dr. Wolf's die was biased and so there is at least one physical property of the die that we have missed that favors some sides more than others. We must search for some traits that differ between the sides.

A standard die is a solid cubical object that is cut as closely as possible into a perfect cube. Small hemispherical depressions ("spots", 1 to 6, depending on the side) are then cut into each face and painted in such a way that the number of spots on opposite sides of the die always[21] sum to 7; i.e. face 1 is always opposite face 6, 2 always opposite 5, and 3 always opposite 4. One obvious way that different sides could differ is therefore in the number of spots inside of their depressions (Figure 4.4b). Here the number of spots on a side is a "trait" that is possessed by all sides of the die but whose value is variable between them just like "height" might be a trait possessed by all species in a plant community but different for different species. So, six depressions are made into side 6 while only one depression is made into side 1 on the opposite side of the die. This would change the center of gravity and shift it slightly towards side 1 since there would be slightly more mass on the 1-face, resulting in side 6 facing up slightly more often. Similarly side 5 would have its center of gravity shifted slightly towards side 2 but a bit less than for side 1 since the difference in depressions on the two opposite sides would be less. Finally, side 4 would be slightly favored over side 3 because the center of gravity would be ever so slightly shifted towards side 3. Notice that we do not need to know the detailed dynamics of the die; we only need to know that the trait could bias the resulting outcome. We can therefore define our first trait for each side as the number of spots on it: $X_{i1} = i$ where i is the number of spots on side i. The observed mean value of our trait is $\overline{X}_{ij} = \sum_{i=1}^{6} i p_i$. The actual average number of dots per throw in Wolf's experiment was 3.5983 and this therefore represents a constraint on the number of ways that one can partition 20 000 throws into six mutually exclusive states. Our two constraints (i.e. the two pieces of information that we now possess) are therefore:

(1) $1 - \sum_{i=1}^{6} p_i = 0$

(2) $3.5983 - \sum_{i=1}^{6} i p_i = 0.$

---

[21] This convention extends into prehistory but no one knows why. Since dice were first used to divine the will of the gods, and since 7 has important magical properties in numerology (remember that Voldemort wanted to split his soul into 7 parts to gain immortality), perhaps this is the origin of the rule?

Table 4.3. *The same information as presented in Table 4.2, plus the predicted proportions (i.e. MaxEnt probabilities, p) given two constraints, and the sign of the prediction errors.*

Side	Frequency	Proportion (o)	Predicted proportion p (constraints 1 + 2)	Sign of error
1	3246	0.162 30	0.152 94	+
2	3449	0.172 45	0.158 18	+
3	2897	0.144 85	0.163 61	−
4	2841	0.142 05	0.169 22	−
5	3635	0.181 75	0.175 02	+
6	3932	0.196 60	0.181 03	+

We maximize the entropy conditional on these two constraints (using the Improved Iterative Scaling algorithm, Box 4.3) and get the following probabilities: $p = \{0.15294, 0.15818, 0.16361, 0.16922, 0.17502, 0.18103\}$. The chi-squared value based on these predicted probabilities is 204.49 with $6 - 2 = 4$ degrees of freedom. This is closer than before, but is this one trait sufficient to explain the experimental results? If it is, then the only remaining deviations between the observed proportions and the predicted probabilities should be due to random sampling variation. This is highly unlikely since the probability of observing a chi-squared value of 204.49 with four degrees of freedom, assuming that only sampling variation is at play, is still astronomically small. We can be confident that we have not yet accounted for all of the sources of bias in our die. This second model includes the constraint of the first model (the second is "nested" within the first) and so the change in the chi-squared value ($269.97 - 204.49 = 65.48$) is also distributed as a chi-squared variate with 1 (i.e. $5 - 4$) degrees of freedom. If this added constraint is irrelevant then the change in the chi-squared value will be due solely to chance but the null probability of observing a chi-squared value of at least 65.48, with one degree of freedom, is 5.6×10^{-16}. Our inclusion of the number of spots, which is linked to a change in the center of gravity, has definitely accounted for some of the bias. Can we think of any other physical features of the sides that might be in play? To get an idea, let's compare the observed and predicted proportions side by side (Table 4.3).

We notice that the sides with 3 or 4 spots tend to have a lower probability of landing up than is predicted by our model with two physical constraints while the sides 1, 2, 5 and 6 tend to appear more often than predicted. Can we think of some physical property of these two sides that might also bias the die? A die is made by clamping a block of material in a milling machine.

Table 4.4. *Shown are the observed number of times and the observed proportions, out of 20 000 throws, that Dr. Wolf's die presented each of the six faces. Also shown is the predicted maximum entropy probabilities given three constraints.*

Side	Frequency	Proportion (o)	Predicted proportion p (constraints 1 + 2 + 3)	Sign of error
1	3246	0.162 30	0.164 33	−
2	3449	0.172 45	0.169 63	+
3	2897	0.144 85	0.141 17	+
4	2841	0.142 05	0.145 72	−
5	3635	0.181 75	0.186 56	−
6	3932	0.196 60	0.192 58	+

Once clamped, the machine cuts the block in the front-to-back and the left-to-right planes. A properly calibrated machine should be able to do this quite accurately. However, to cut the last plane (top–bottom) the block must be removed from the machine, turned upside down, and the machine again adjusted to make the final cut. This action might result in the cube being slightly oblate, with one dimension slightly shorter than the other two, or being slightly prolate, with this dimension being slightly longer than the other two (Figure 4.2c). Judging from the values in Table 4.3, it appears as if the opposite sides (faces 3 and 4) have a slightly lower probability of occurring than the other two dimensions and so the die might have been slightly prolate with the 3–4 dimension being slightly greater than the other two, thus again affecting the center of gravity. Suppose that the 3–4 dimension was greater by some small amount δ resulting in the other two dimensions (that are equal) being favored by an amount $k\delta$. We have no way of knowing the value of $k\delta$ but this isn't necessary. We only have to assign a second trait whose values are proportional to it. We define trait 2 as: $t_2 = \{1, 1, 0, 0, 1, 1\}$. We could have used $t_2 = \{0, 0, 1, 1, 0, 0\}$ or $t_2 = \{10, 10, 2, 2, 10, 10\}$, or any other coding that gives equal weight to sides 1, 2, 5 and 6, and different but equal weights to sides 3 and 4. The only effect of these different weights would be to change the λ values while keeping the proportions the same. Using the observed proportions in the experiment, the average value of t_2 is 0.7131; this is our third constraint. The result is shown in Table 4.4.

The chi-squared value for this third model is 9.38 with $6 - 3 = 3$ degrees of freedom. The probability that the differences between the observed and predicted proportions are due only to sampling variation is 0.03. The change in chi-square relative to model 2 is $204.49 - 9.38 = 195.11$ with one degree of freedom. The null probability of this change in chi-square is again very small, meaning that our second trait has again improved the predictive ability of our model. It appears that we still have not accounted for all of the bias but we are almost there! In Table 4.4 we see that the sides 2, 3 and 6 still have a slight positive bias. What physical feature of a die could account for this? Looking at a real die (Figure 4.2d) we see that these three sides come together at one corner. What if there was a small chip missing from this corner? This would give a tiny positive advantage to these three sides relative to the other three. We can therefore propose another trait: $t_3=\{0, 1, 1, 0, 0, 1\}$ to represent this hypothesized bias. Maximizing the entropy with respect to all three traits, we get a chi-squared value of 0.79 with two degrees of freedom. The probability of observing this degree of difference between observed and predicted proportions, if sampling variation was the only remaining cause, is 0.67. The change in the chi-squared value due to adding this third trait is $9.38 - 0.79 = 8.59$ (change in degrees of freedom is still one). Thus, it appears that these three physical features of the die are sufficient to account for all of the bias, at least based on the amount of statistical power that 20 000 independent throws provides. Once these three traits are taken into account, the six sides are equivalent, or neutral,[22] with respect to any advantages they have and any remaining differences are due solely to sampling variation.

Here is the final model:

$$p_i = \frac{e^{0.0303t_{i1}+0.2149t_{i2}+0.0424t_{i3}}}{7.9169}.$$

Looking at each of the Lagrangian multipliers (the λ values) we can calculate by how much a change in one unit of a trait will change the proportional relative abundance of each side if all other traits don't change. For trait 1 (the number of spots per side) we have: $-\lambda_1 = \partial p/p\partial t_1 = 0.0303$. Thus adding one extra spot increases the chances that that side will land up by 3%. For trait 2 (being in the 3–4 dimension of the cube) we have $-\lambda_2 = \partial p/p\partial t_2 = 0.2149$. Being in this dimension increases the chances that the side 3 or 4 will land up by 21%. For trait 3 (the chip out of the 2–3–6 corner) we have: $-\lambda_3 = \partial p/p\partial t_3 = 0.0427$. The chip increases the chances that sides 2, 3 or 6 will land up by 4%. Applying the analogy with an ecological community, we could say that these λ values quantify the selective advantage of possessing the associated trait.

Let's consider what we have achieved. The explanatory variables in this simple model are physical attributes of the different sides of the die. We can

---

[22] I use the word " neutral" to point out the similarity to neutral models of community assembly.

provide a physical (mechanistic) explanation for why each variable would bias the relative abundance of throws in favor of each face. We can therefore link certain physical forces acting on individual throws to a collective property (the relative abundance distribution which is the macrostate). Of course, we can't perform an experimental test of this mechanistic link in the case of Dr. Wolf's die because it is long gone, but we could easily do this with a contemporary one. Although the explanatory content of the model is mechanistic, the mathematical link is not; we did not produce our predictions by modeling the detailed dynamics of each throw as it left Dr. Wolf's hand. On the contrary, our predictions were obtained by *ignoring* both the detailed dynamics and the idiosyncratic initial differences of each throw and concentrated on those aspects of each throw that were both reproducible and likely to bias the outcome of a particular throw in a consistent way. And yet, despite the small amount of empirical information that we used to construct our model (three attributes of the six sides plus the average values of these three attributes) we were able to predict the observed relative abundances highly accurately in a data set consisting of 20 000 independent observations. Even with a sample size that is at least an order of magnitude larger than most ecological data sets, and with a resulting level of statistical power that can detect even very small systematic deviations between observations and predictions, we could account for all non-random patterns. We will use this same strategy to link organismal traits to ecological communities.

The components of a Maximum Entropy model

Now that we have learnt the mechanics of the Maximum Entropy Formalism and seen a small example of how it works in practice, let's decompose it into its basic parts.

Entity. The basic unit of our model is an entity that can exist in different states. The total number (N) of such entities are arranged in time (thus producing a temporal dynamic) or space (thus producing a spatial dynamic). If our system refers to a collection of N die that are thrown together then each die (a distinct entity) is differentiated by its spatial position. If our system is a single die that is thrown N times in succession then each throw (a distinct entity) is differentiated by its temporal order. If our system refers to a collection of individual plants (entities) existing at a site (thus N individuals in space) then each entity is a single individual. If our system refers to a collection of molecules (entities) existing within all of the living plants at a site (thus N molecules of biomass in space) then each entity is a single molecule.

If our system refers to a collection of larger biomass units (whatever these might be) existing within all of the living plants at a site then each entity is this larger biomass unit.

State. Once we decide upon the basic entities of the model, we must then decide how we want to classify the different, and mutually exclusive, ways in which a single entity can exist. If our system refers to a collection of throws of a die then each entity (a throw of a die) can be classified by the number of different faces and each unique face is a state. For a collection of individual plants or units of biomass then each entity can be classified by the species to which it belongs and each unique species is a state. If our system refers to a collection of species and each species can be classified by its abundance (number of individuals, amount of biomass . . .) then each species is an entity and each unique abundance value is a state.

Perhaps the easiest way of identifying what constitutes an entity, a state and (below) a trait of an entity, is to draw a bivariate graph of the eventual output of your model. The abscissa will show the probability, or relative abundance, of *what*? The "what" will define your entities. The ordinate will show the way you have classified your entities. Once you label the ordinate then you will have defined your states. Now, if there exists some property that is common to all entities, but whose value differs between entities existing in different states, then this defines a trait. Figure 4.5 shows some different possibilities. Figure 4.5a plots the relative abundance of individuals found in five different species. Here, the entity is a single individual, the states are the different species into which one can classify the individual. If entities (individuals) in different states (species) vary in some trait (height, longevity . . .) then these define possible traits for the model. Figure 4.5b plots the relative abundance of biomass found in five different species. Here, the entity is a single unit of biomass and the states are the different species into which one can classify a single unit of biomass. Figure 4.5c plots the proportion of species (i.e. a relative abundance of species) having given amounts of biomass. Here, the entity is a species and the states are (continuous) biomass amounts.

Note that, in Figure 4.5a, each entity is easy to count and so the value for N is easy to obtain. In Figure 4.5b it is not obvious what constitutes the "units" of biomass that are distributed in time or space. If we choose to measure biomass in grams then we would get one value for N but someone else could use kilograms and get a different value for N. This is the problem of arbitrary units that our maximally uninformative prior will have to address but notice that we could have had the same problem in Figure 4.5a. If we define a group of 100 individuals as a unit of *gloop* and a group of 10 individuals as a *centigloop* then our value for N would also change in Figure 4.5a depending

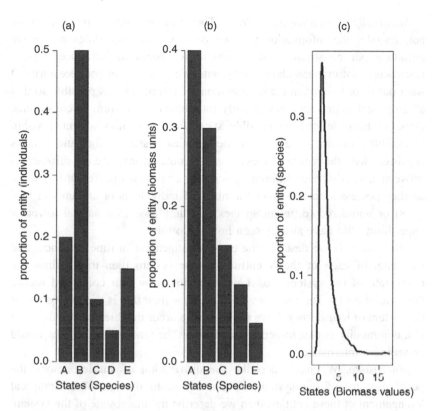

Figure 4.5. Three different ways in which entities can be classified into states.

on whether we measure in gloops or centigloops. So, our value of N (and thus, our definition of an entity) must refer to the smallest natural division of the things that are *independently* allocated to different spatial or temporal positions; these are the degrees of freedom of our system (not of our model). Unlike simple systems like throws of a die, it is not always possible to unambiguously determine these degrees of freedom but, as you saw earlier in this chapter, we don't actually need to know the value of N in order to maximize the entropy. None the less, when there are temporal or spatial correlations between the states of entities that are expressed, then statistical inferences about significant deviations between observed and predicted probabilities or relative abundances becomes more difficult.[23]

---

[23] This problem can easily occur even with more obvious entities like "individuals" when we are dealing with plants when different aboveground stems (ramets) are connected (or where connected in the past) by underground rhizomes.

Maximally Uninformative[24] Prior. The prior (maximally informative or not) encodes the information that we possess about our states before we introduce our constraints and so before we begin to introduce explicit macroscopic information about our system. We usually (but not necessarily[25]) want our prior to contain the least amount of information as possible so that all empirical information is explicitly introduced in the form of constraints. Once we have defined the possible states for our entities we must assign probabilities to each state that encode the least possible information that is consistent with the nature of these states. In other words, are the differences between states discrete (different species) or continuous (different biomass), do they possess a natural order (numbers of individuals or amounts of biomass) or unordered (different species), do they possess a natural lower or upper limit? We have already seen how to do this.

Microstate. If we describe the exact arrangement in time or space, and the state, of each of the N entities in our system then this defines the microstate of the system. So, if we were to throw four coins and record the state of each – for instance, $\{H, T, T, H\}$ – then this is the microstate of the system of four throws. If we are to see the order in which each individual in a community comes to occupy a space on the landscape, then we would be seeing a microstate.

Macrostate. When we describe the distribution of entities among the possible states of our system without regard to the spatial or temporal arrangement of these entities, then we describe the macrostate of the system. If we were told that the outcome of throwing four coins was two heads and two tails, but not the spatial or temporal order of the heads and tails, then this is the macrostate of our system. If we are observing the outcome of allocating biomass to different species (a relative biomass distribution) but not the physical process of atoms being captured by individuals of different species, then we are observing a macrostate. Figure 4.5 shows macrostates, not microstates. Since the whole point of a MaxEnt model is to predict the macrostate without having to (or being able to) know the detailed dynamics of how the system has developed in time or space (i.e. to be able to predict the microstate), the notion of a macrostate is important.

---

[24] An equivalent term is an ignorance prior. Both "maximally uniformative" and "ignorance" refer to the fact that these prior probabilities reflect what we know about these states purely from the way they are defined and in the absence of any empirical information; i.e. what we know before Nature gives us any information.

[25] In particular, since we are ultimately interested in constructing a model representing trait-based environmental filtering, we might want to include information on other processes, not related to traits, into the prior. We will do this, for example, in order to express a prior assuming neutral community dynamics.

Traits, attributes or properties of our entities. In general, each entity will also possess measurable properties whose values differ between states. So, if different faces of a die have different numbers of holes, or different surface areas, or different centers of gravity, then these are traits associated with each state. If different individual plants have different heights, or produce different numbers of seeds, then these are traits associated with each state.

Traits, attributes or properties of macrostates. The traits of entities, when averaged over an entire macrostate, define an aggregate property. So, the average number of heads in a series of throws of a coin is a macrostate trait. The average seed production per individual plant is a macrostate trait of the vegetation of a site. An equivalent term in ecology is a "community-aggregated" trait. Such macrostate traits define the empirical constraints whose values we must respect when maximizing the (relative) entropy and thus determining the probabilities assigned to each state and the most probable macrostate.

Inferential statistics of a MaxEnt model

Once we have obtained the posterior probabilities (proportions, relative abundances) of our MaxEnt model given certain constraints, we will usually want to ask one (or all three) of the following questions. First, does our resulting model, including all of the constraints, agree with Nature? Second, does a proposed constraint (i.e. a community-aggregated trait) actually affect community structure and thus relative abundances? Third, how well does our model predict the actual community structure? As we will see, the first question can be rigorously answered under some conditions but not under others while the last two can always be answered.

Question 1: does our resulting model agree with Nature; that is, does the empirical information that we have included in the model in the form of constraints account for all of the observed differences in the actual proportions of entities expressing different states? In essence this first question requires us to test a null hypothesis that the observed and predicted proportions are the same except for the deviations that one would normally expect to arise simply from randomly allocating N mutually independent entities in K groups (states) while constraining these allocations to agree with $T + 1$ pieces of information. Here T is the number of macroscopic (community-aggregated) traits but we always add one extra constraint in order to insure that the sum of the predicted proportions is unity. Given the observed (o) and predicted (**p**) proportions in each of K states, the number (N) of mutually independent entities (N), and the number (T) of macroscopic traits, the following statistic is distributed as a chi-squared deviate with $N-T-1$ degrees of freedom if the observed differences between observed and predicted proportions are due solely to sampling variation: $G^2 = 2N \sum_{j=1}^{K} o_j \ln(o_j/p_j)$. This, of course, can be equivalently calculated as $G^2 = 2 \sum_{j=1}^{K} n_j \ln(n_j/\hat{n}_j)$, where $\hat{n}_j$ is the predicted number of entities, by converting from proportions to the actual number (n_j) of observed entities in each group.

There are some rather important conditions that must hold in order to use these equations. One must either know N (the number of *independent* entities that have been allocated to the states) and the proportions (p_j) or else know the actual numbers (n_j) of entities in each group in order to conduct the test. In simple cases, such as Dr. Wolf's die experiment, this is not difficult but in many (perhaps most) ecological communities this is much more difficult. When the states are discrete, like the number of individuals per species or the number of species having a given number of individuals, it is at least feasible to make such counts but it is much more difficult to ensure independence. In many plant species more than one physical individual (a ramet) arises from the same genetic individual (a genet). Furthermore, the spatially restricted nature of vegetative and sexual reproduction means that there will be spatial and temporal correlations between the states expressed by different entities. An even more difficult situation is when the states are continuous, like biomass. What would be the value of N in such situations? If each entity is a single atom then N would be the total number of moles of atoms in the biomass – an incredibly large number. But plants don't take up atoms independently of one another. Plants must respect stoichiometric constraints meaning the N would be much smaller than the total number of moles of atoms. Furthermore, the spatial and temporal correlations between resources again reduce the number of *independent* events.

All of these considerations suggest to me that this first question cannot be answered in a rigorous manner in any real ecological problem. At best one might turn the question on its head and ask instead what the value of N would have to be in order for the null hypothesis to be accepted at a specified significance level. More generally, I strongly suspect that no MaxEnt model, tested against a reasonably large ecological data set, would fit perfectly. This is because there are always a very large number of causal factors affecting community assembly and we are unlikely to ever be able to measure (or even know about) all of them. Furthermore, some causal factors in any local community will likely be idiosyncratic properties, perhaps related to the specific history of the site, that are not generalizable and so are of no real interest.

Second question: does the addition of a specified constraint into a previous model improve our ability to predict the observed proportions (relative abundances)? This second question requires us to test the null hypothesis that there is no *change* in the overall deviations between observed and predicted proportions, once we include the added constraint, beyond what one would expect from sampling variation.

Let's first consider the less common situation in which the conditions applicable to our first question hold. If G_T^2 is the G^2 value of a MaxEnt model consisting of T macroscopic constraints and G_{T+1}^2 is the G^2 value of a MaxEnt model obtained from the same data set and consisting of the same T macroscopic constraints plus one extra then, if the added macroscopic trait is independent of the observed proportions, the difference $(G_T^2 - G_{T+1}^2)$ will follow a chi-squared distribution with one degree of freedom. This is the well-known property of nested models.[26]

For the reasons explained above, the conditions necessary to apply the G^2 test are limited in ecological situations. Often we know the relative abundances, for instance when measuring abundance by biomass or some other continuous measurement of abundance, but we do not know N. Other times, even if we can count each entity, we are quite sure that temporal and spatial correlations between the entities exist, thus reducing the number of independent events. Given these considerations, it is probably more useful, and tractable, for ecologists to instead ask the following two questions: (1) how well does our model predict the observed relative abundances; and (2) how much better is the predictive ability of our model if we add a new community-aggregated constraint to our model?

Stated in this way, our problem is one that is common to statistical modeling. We need to quantify the degree of association between observed and predicted proportions, determine the probability of such an association arising by chance (thus, does our model have any predictive ability?) and then determine the probability of such an association increasing by our observed degree once we add a new explanatory variable (i.e. a new community-aggregated trait) to our model.

A possible parametric solution, and one that can be implemented in virtually any statistical program, is to calculate a Pearson coefficient of determination (R^2)

---

[26] In general, if we add X additional macroscopic constraints to a previous model then the difference, $G_T^2 - G_{T+X}^2$, will follow a chi-squared distribution with X degrees of freedom.

between the observed and predicted proportions. Of course, the R^2 statistic is a measure of the correlation between observed and predicted values, not of the agreement between the two; for that one would use a residual mean square. In order to draw statistical inferences one must insure that the proportions are linearly related and that the observed proportions are approximately normally distributed, and this will rarely be the case. When this is the case an inferential test for such an R^2 statistic, based on N_p observed proportions and K explanatory variables (i.e. community-aggregated traits) is obtained by calculating Fisher's F ratio:

$$F_{K,\, N_p - K - 1} = \left(\frac{R^2}{1 - R^2}\right)\left(\frac{N_p - K - 1}{K}\right).$$

Now, in order to test whether or not the addition of X new community-aggregated traits has significantly improved the fit between observed and predicted proportions one calculates the R^2 values for the first model (R_1^2) based on K_1 community-aggregated traits and for the second model (R_2^2) based on $K_1 + X = K_2$ community-aggregated traits:

$$F_{K_2 - K_1, N_p - K_2 - 1} = \left(\frac{R_2^2 - R_1^2}{1 - R_2^2}\right)\left(\frac{N_p - K_2 - 1}{K_2 - K_1}\right).$$

The degrees of freedom given in these equations assume that one is using the same data to both estimate the community-aggregated traits, thus estimating K parameters, and then calculate the predicted relative abundances. If you can predict the community-aggregated traits from independent data (from environmental conditions for instance) then you would not loose these degrees of freedom. Of course, one must always fix one constraint in order to insure that the relative abundances (predicted probabilities) sum to unity, so we always loose at least one degree of freedom. This is exactly analogous to what one does in multiple regression: one first uses the data to estimate the parameters in the regression model (i.e. the partial slopes and the intercept) and then one compares observed and predicted values based on the residual degrees of freedom.

The application of this inferential statistic must be made with care until proper simulation studies have been conducted. My own very preliminary simulations suggest that the Pearson coefficient of determination does not always perform very well but that replacing it with the Spearman coefficient of determination[27] in the calculation of the F ratio might give reasonable

---

[27] This is simply the Pearson R^2 between the rank-transformed observed and predicted values.

probability levels, at least at the most common significance levels (i.e. $p = 0.2$, 0.1, 0.05 and 0.025). My preliminary simulations suggest that even skewed or binary traits do not affect this conclusion much. Similarly, the presence of many "missing" species (species that are in the species pool but are not present in the local community) did not greatly affect the result. Of course, there is much more work to do before any of these conclusions can be generally applied. However, if you want more precise probability levels then it is best to use permutation methods, as suggested by Roxburgh and Mokany (2007). To test the significance of the traits in explaining differences in relative abundance one would first calculate the F ratio for the actual data. Next, you randomly permute the rows of the species (rows) X traits (columns) matrix; doing so maintains the actual patterns of covariance between traits but forces these traits to be independent of the observed relative abundances. You would then calculate the community-aggregated trait values given these permuted trait vectors, re-fit the model and calculate the permuted version of the F ratio (or any other appropriate fit statistic). After repeating this many times you would then count the proportion of times that these permuted F ratios exceed the observed F ratio. This proportion is a permutation estimate of the probability of observing such a degree of predictive value by chance given the null hypothesis that the traits are independent of the relative abundances. If you want to test the null hypothesis that only trait X (say) was independent of the relative abundances, then you would randomly permute just this one trait across species, and and compare the change in R^2 values based on the second F ratio given above. However, such permutation methods are computationally very expensive, especially when you must repeat this across many different local communities, because the precision of the permutation probability increases with the number of random permutation runs; at least 1000 are needed for a 5% significance level (Manly 1997).

Finally, we would often want to conduct an inferential test based on fitted values from many different local communities; a "local community" will often coincide with a "quadrat". If these local communities are sufficiently far apart in space that they can reasonably be considered independent then we can combine the probabilities obtained from testing the fit in each local community (let's call these p_i) by assuming that traits are independent of predicted relative abundances *in general*. If this is true then such p_i values, obtained from Q independent local communities, will follow a uniform distribution under the null hypothesis and the following statistic will be distributed as a chi-squared variate with $2Q$ degrees of freedom (Fisher 1925) and unusually large values will be evidence to reject the null hypothesis and conclude that traits are related to relative abundances:

$$-2 \sum_{i=1}^{0} \ln(p_1) \sim \chi_{2Q}^2.$$

A word of caution. Although known since early in the twentieth century (Fisher 1928a, 1928b), ecologists continue to be surprised (Roxburgh and Mokany 2007) to learn that, even when predicting a variable (Y) from a set of mutually independent variables ($X_1, X_2 \ldots$), the expected R^2 value will not be zero (i.e. it is biased away from zero) unless one is working with an infinite sample size. This can be seen intuitively by answering a series of simple questions; let's use the simplest case of one dependent variable (Y) and one independent variable (X). First, assuming X and Y are independent, what is the expected value of Pearson's r statistic? It is zero so long as we have at least three observations per data set;[28] some data sets would result (by chance) in negative r values and some would result (by chance) in positive values with the mean being zero. Second, how far will the r values of particular data sets deviate from this expected value of zero? This will depend on the sample size. Small sample sizes will result in a wide spread around zero and large sample sizes will result in a narrow spread around zero. Third, how will the answers to the first two questions change when we square our values of r to get r^2? Well, from our answer to question one, we deduce that all r^2 values will be greater than zero, so the expected value of the r^2 values will always be greater than zero. From our answer to the second question we deduce that the expected value of the r^2 values will be close to zero for very large sample sizes and far from zero (i.e. closer to 1 and therefore more biased) for small sample sizes. In fact, if we only have two observations per data set then our expected r^2 will be 1! This is why the degrees of freedom in testing for a correlation in this simple bivariate case is the number of observations in the data set minus 2. We can calculate a bias-corrected value of R^2 using the following formula:

$$R_{adj}^2 = 1 - \left(\frac{N_p - 1}{N_p - K - 1} \right) (1 - R^2).$$

Avoiding circularity

A number of papers (Marks and Muller-Landau 2007, Roxburgh and Mokany 2007, Haegeman and Loreau 2008) have criticized the predictions of the

---

[28] If only two data points, then the expected value is 1 since one can always fit a line that passes exactly between any two points.

MaxEnt model as being "circular". It is time to address this criticism directly because it is related to the inferential statistics explained above. Unfortunately, it is not clear to me how I should do this because none of these papers defines what is meant by "circular". From what I can tell the charge of circularity involves two components. The first component seems to be the idea that the model is running in circles: we predict relative abundances from community-aggregated traits but we need to know the relative abundances in order to calculate the community-aggregated traits. Isn't this like a dog chasing its tail? The second component seems to be that the predictive ability is reflecting some necessary mathematical consequence of the model rather than telling us anything about how Nature behaves.

Let's start with the first component, since this is easier to explain. In the full model, we would predict the community-aggregated traits based on empirical relationships between such community-aggregated traits and environmental gradients, as explained in Chapter 3. We would then use these predicted community-aggregated traits, given the environmental conditions, not the *measured* community-aggregated traits, in the MaxEnt model. In the full model the dog is running in a straight line, not in a circle.

However, it is perfectly valid to also use the measured community-aggregated traits as well. In fact, in Jayne's classic analysis of Dr. Wolf's die that was described earlier we did the exact same thing. In the die example we used the frequencies from the 20 000 throws of the die to calculate the average (i.e. "aggregated") values of the "traits" and then used these to predict the observed frequencies. In classical statistical mechanics, which can be formulated as a MaxEnt model (Jaynes 1957a, b, 1983, 2003), one uses the measured temperature of a gas to predict the distribution of energy states of the atoms. But "temperature" is simply the average kinetic energy of these atoms; "temperature" is a measured aggregated value of this community of atoms. Of course the temperature is measured directly, not based on a weighting by frequencies, but one can also measure community-aggregated traits directly rather than calculating them from relative abundances (Gaucherand and Lavorel 2007). This is done by randomly sampling individuals, irrespective of the species to which they belong, and taking the arithmetic average of the measured traits. Why we are justified in doing this requires a discussion of the residual degrees of freedom of a model, which will be explained in the next section.

For the moment, and simply to convince you that the MaxEnt model (using observed community-aggregated traits) is not the prestidigitation of a conjurer, I will simply point out that you have probably done the same thing many times. You aren't a sham artist, are you? Consider the following case of a simple bivariate linear regression. You have a data set with N independent observations of two variables (X and Y). You use this data set to estimate the parameters of the linear regression (the slope and the intercept) and this is done by minimizing the observed and predicted values of Y (the residual sum of squares). You then use the *same* data to calculate the goodness of fit (a correlation coefficient or a residual mean square) and to test the statistical significance of this fit. In practice these two steps appear to occur simultaneously in your computer but in fact they are sequential. In other words you predict the values of Y given the values of X, but you (or your computer) must already know the values of Y (and X) in order to do this. The dog is again chasing its tail! In both the MaxEnt model (using measured community-aggregated traits) and in this simple regression example we must already know the values of Y in order to predict them. In both cases we are fitting the model and testing it with the same data.

A more precise explanation of why the dog is not really chasing its tail, either in regression or in MaxEnt models, requires that we now consider the second component of the charge of "circularity". What does this second component mean when claiming that a model prediction is circular? It means that the model output (the predictions) is entirely determined by the logical constraints of the model, not by the relationship between the model and Nature, and so the model predictions are simply a tautological restatement of the model input. It is like saying "my father is a man". It means, as I will later explain, that there are no residual degrees of freedom.

Consider again the simple bivariate linear regression. To specify this simple model we need to know the values of the slope and intercept. To evaluate the relationship between the model output and Nature (i.e. its predictive ability) we need to measure the deviation of observed and predicted values, given the slope and intercept. If we can state the amount of deviation between observed and predicted values without even having to look at the observations – that is, without even asking Nature – then this predictive ability is independent of Nature and is entirely conditional on the mechanics of the model (here, values of the slope and intercept). The model predictions would be circular because the model output (the degree of deviation) would be entirely dependent on the model input (the slope and intercept). How might this occur?

Choose any two points (sets of X–Y coordinates) and place them on a graph. Don't tell me how you decided to chose these two points; they could come from some population in which X and Y are independent, or moderately correlated, or even perfectly correlated. I don't need to know. It doesn't matter. The degree of deviation between observed and predicted values will be zero. Why? Intuitively, the answer is because a least squares solution chooses the slope and intercept by minimizing the deviations between the observed and predicted values and we can always draw a straight line between any two points. It is logically impossible for there to be any deviations from such a line irrespective of how X and Y actually behave in Nature. The perfect fit (zero deviations) is independent of Nature and is entirely dependent on the mechanics of the least squares algorithm. Because of this, we cannot know if the predicted values in Nature are good or bad. The model prediction is entirely circular. On the other hand, if you choose any three points from your population then I cannot possibly tell you what the degree of deviation will be. Once we have three or more independent observations then the model predictions are not dependent on the logic of the model (the least squares algorithm); rather, they are dependent on Nature, and so the model predictions are no longer circular.

This notion can be made more precise and is explained in detail in Mulaik (2001); what follows is a less technical version of Mulaik's argument. If you have N independent observations from Nature then there are N degrees of freedom for these data; there are N independent "bits" of information. These are the total degrees of freedom. More precisely, we need N dimensions in order to fully specify where the dependent values of our observations actually lie within the space of where they might logically lie. If your model needs the values of K parameters, and if you use the observations to estimate these values, then you have used up K "bits" of information that are contained in your data. These are the K model degrees of freedom. More precisely, we need K dimensions in order to fully specify where the parameters lie within the space of where they might logically lie.

The residual degrees of freedom of the data (v_R) is the difference between the degrees of freedom of the data (N) and the degrees of freedom of the model (v_M): $v_R = N - v_M$. The residual degrees of freedom of the data are the number of "bits" of information contained in the data that have not been used up to determine the model parameters; Nature is "free" to chose their values independently of the constraints placed by the model. In the case of a linear regression with only two points we start with $N = 2$ dimensions in which the two Y values lie. Once we use these points to estimate the slope and intercept then, given this information, we have reduced the number of dimensions that we need to locate these points by 2. The number of dimensions (the residual number of dimensions) that we would still need to locate our two points, given our parameter values, is $v_R = N - v_M = 2 - 2 = 0$. In other words, the location of the points is now fixed (they are no longer free to vary) and we can say exactly where they are without even asking Nature. In the case of the linear regression with three points the residual number of dimensions that we would still need to locate our two points, given our parameter values (the residual number of dimensions), is $v_R = N - v_M = 3 - 2 = 1$. Because this remaining dimension can't be obtained simply from the mechanics of the least squares algorithm, the location of our data set along this remaining dimension is not fixed by the mechanics of the model and so it has to be observed in Nature. It is precisely *because* the value must be observed, rather than derived, that allows us to test[29] the predictive ability of the model because Nature can now disagree with the model. The more residual dimensions (i.e. the more residual degrees of freedom) that exist, the better we can determine if (and to what extent) Nature really agrees with our model.[30]

As long as there are residual degrees of freedom the model is not circular. This is why, in conducting inferential tests of fit for a model, one uses the residual degrees of freedom. The residual degrees of freedom in a simple bivariate linear regression that is fit to empirical data is $N - 2$ because we need to use the data to estimate two parameters (the slope and intercept). The residual degrees of freedom of a multiple linear regression with V predictor variables is $N - V - 1$ because we need to use the data to estimate $V + 1$ parameters (an intercept and V partial slopes).

Now, let's look at our MaxEnt model. We must estimate values for each of the model parameters (the Lagrange Multipliers) in order to fit the model. There are $T + 1$ parameters because we have $T + 1$ constraints (the normalization constraint plus the T community-aggregated traits). We need to use up $T + 1$ "bits" of information in our empirical data to estimate these parameters if we estimate them from our data rather than specifying them from independent information (from a knowledge of environmental gradients, for example). The total number of dimensions over which the data can vary is equal to the number of states (S) and this is the number of species in the species pool. The residual degrees of freedom of the model is therefore $S - T - 1$. If our model has S species in the species pool and $T - 1$ traits then the model is circular because the predictive ability of the model is independent of Nature. It is like fitting a regression to two points. As long as the number of species in the species pool is greater than $T + 1$ then the model is not circular.

---

[29] Of course, with only one free dimension for such a test, it is not a very powerful one.

[30] Residual "degrees of freedom indicate the number of dimensions that still remain in which the data is free to differ from the model's best representation of the data, and thus indicates the potential disconfirmability of the model. The greater the number of ways (dimensions in which) the hypothesis could be disconfirmed, the greater the confidence in the objective validity of the model, if it fits the data well." (Mulaik 2001). Here "disconfirmability" means "falsifiability", that is, the ability of the data to deviate from the model and thus to disagree with it.

Take the example of a simple coin toss which exists in two states (heads $= 1$, tails $= 0$). If I tell you that the average value from many tosses was 0.45, then can you tell me the distribution of states (heads and tails) that actually occurred in this particular data set without actually looking at it? Of course you can. The answer (head/tails) is 0.45 and 0.55. You could do this because there were only two states and you knew the values of two con-straints: $p_H + p_T = 1$ and $1 \cdot p_H + 0 \cdot p_T = 0.45$. Stated equivalently, there are two total degrees of freedom in the data and you must estimate two param-eters in the MaxEnt model. The residual degrees of freedom are zero. The answer is completely fixed by the two constraints (the two pieces of empirical information) and so it is impossible for the data to disagree with the answer. What if we had a 6-sided die, like the one used by Dr. Wolf, and I told you only that the average face value of many throws was 3.8. Could you tell me the distribution of states (the six sides) without actually looking at the data? If you can then you know a lot more about this die than I have told you. All you really know are that $p_1 + p_2 + \cdots + p_6 = 1$ and that $1p_1 + 2p_2 + \cdots + 6p_6 = 3.8$. Since this is not enough information to determine the entire dis-tribution, therefore there are many different distributions that could occur. The residual degrees of freedom $(6 - 2 = 4)$ are greater than zero. If you choose the wrong one then Nature will disagree with you and your model must be rejected. Because the predictions are potentially falsifiable, the model is not circular.

A "statistical mechanistic" model of community assembly

Here, then, is a description of the Maximum Entropy Formalism, applied to plant communities, that is suitable for a T-shirt: *find the distribution of relative abundances of species from a species pool that maximizes the relative entropy subject to community-aggregated trait values*. From a purely infor-mational point of view this is simply an acknowledgement of ignorance. The only information that we possess about the detailed dynamics leading to the state (the species) expressed by a particular entity (the individual or unit of biomass) are encoded in our constraint equations (community-aggregated traits). Since this is all we know, we must admit to our ignorance (i.e. choose maximal uncertainty) after taking into account these constraint equations. Stated thus, our model is simply a prediction device – potentially useful but not biologically insightful. In order to provide explanatory power to our model we must show that equations of the form $\sum_i p_i X_{ij} = \overline{X}_j$, i.e. community-aggregated traits, are logically implied given known biological processes related to community assembly. This is the topic of Chapter 5.

From a mechanistic point of view the maximum entropy formalism is both an acknowledgement of ignorance and a claim of irrelevance at the macroscopic (community) level. We are assuming that the underlying causal forces that determine the population dynamics of the species during community assembly, and that therefore causes a particular individual or unit of biomass to belong to a given species, are of two types. Those causes that are encoded in our constraint equations are repeatable over time and space and so are reflected in the resulting community structure. All other causes determining why a particular unit of biomass belongs to a given species are idiosyncratic, their outcomes show no repeatable patterns over time and space, and so such causes do not bias the outcomes in favor of any particular state.

If we think again of Dr. Wolf's die, there were certain causal forces that acted consistently from throw to throw: those affecting the center of gravity and determined by the number of holes per face, the relative lengths of the sides and the chip in one corner.[31] There were many other causal forces affecting the dynamics of particular throws, such as the force with which Dr. Wolf threw the die on the 23rd throw, the angle of the 567th throw, the amount of angular momentum in the 2435th throw, the door that opened on the 1985th throw causing a gust of wind and so on, but these were not repeatable from throw to throw and so did not result in a consistent bias[32] (Mosekilde 1996). We know, since the deviation of his observed distribution from a uniform distribution was actually very small, that these idiosyncratic causes were more important than the repeatable ones, leaving the slight repeatable biases to emerge.

We can imagine the same thing happening during community assembly. Think of a local species pool of S species (i.e. all those species physically able to reach a site) as a biased die with S sides. Now, what sequence of events must happen before a plant ecologist records that a particular individual of plant species i occupies point X at a given date? First, species i must be present in the site in order for one of its propagules to reach point X and this means that one of its propagules (the individual in question or one of its

---

[31] This, as far as I can tell, is what Popper really meant when he talked about "propensity" probabilities: those physical causes of an event that have consistent effects in repeated trials and so bias frequencies of occurrence.

[32] "Unpredictability [in a die toss] enters through such phenomena as scratches and inhomogeneities in the table and through the inability of the tosser to control the initial conditions well enough. For higher tossing frequencies where the mixing [of dynamic trajectories] becomes stronger, vibrations of the table, airflows, gravitational interactions of the die with nearby bodies, and at the end even thermodynamic fluctuations may influence the outcome" (Mosekilde 1996, page 45). A detailed description of the nonlinear dynamics of the throw of a die is given in the chapter "How random is a die toss?".

ancestors) must have succeeded in immigrating to the local site from some other population. Such long-distance dispersal depends on a number of contingent, and idiosyncratic, factors (the right gust of wind or the right animal vector coming at the right time). Such contingent factors will not be repeated consistently over time and space, but some repeatable properties of propagules (seed size, the presence of barbs on the seed, seeds inside a digestible fruit) will bias the chances of certain species succeeding in immigrating to the site. Once a propagule of species i arrives at point X it must survive predation and fungal, bacterial and viral attack, perhaps for a long time if it resided in the soil seed bank. Such events will depend on a complex sequence of causes that will vary from point to point and over time but, again, certain repeatable properties of the seed in combination with the surrounding environment will bias the chances of a given species surviving until germination. The proper germination cues must then present themselves and the immediate abiotic conditions, themselves modified by the presence of other plants, must favor our individual of species i surviving to reproduction rather than one of its neighbors. Finally, our individual must successfully flower, set seed and disperse its seed. Each step of the life cycle requires a complex combination of causes to combine in just the right way for point X to be occupied by an individual of species i rather than by species j. If, rather than thinking of individuals occupying space, we think of atoms being captured and incorporated into biomass, then the complexity with which causes interact is even greater.[33]

Described like this, we almost have the impression that the structure and dynamics of plant communities are the outcome of a giant game of crooked craps played by Nature. Each second there are a huge number of biased dice being thrown with the state (the species) that each entity (individual or atom of biomass) comes to express being determined by the throw of these biased dice.[34] Like Dr. Wolf's die that had certain traits whose repeatable causal forces biased the outcome (however slightly), the dice that Nature throws are also biased such that, over many repetitions, certain outcomes become more likely than others by virtue of certain organismal traits in a given environmental context. Imagine that we had a die with as many faces as there are species in the potential species pool[35] (i.e. all those species for which it is

---

[33] This, incidentally, is why Banquo's challenge, cited in the preface, is likely never to be met.
[34] "God plays dice with the universe, but they're loaded dice." Attributed to the physicist Joseph Ford in response to Einstein's famous quote that "God does not play dice" (Gleick 1987, page 314).
[35] Of course, it is only physically possible to have certain numbers of faces per die and it is not easy to work out from first principles which numbers are possible. The solution is given in an MSc thesis by Taylor Pegg, Jr (Taylor Pegg Jr 1994).

physically possible to reach the site in question). A single throw of this die is analogous to a single atom from this site being captured by a species and converted into living biomass. The total number of atoms in the biomass of the site is the result of a huge number of throws of the die (although certainly not mutually independent throws). We can't see the actual sequence of throws (i.e. the microstate dynamics consisting of the sequence of atoms being captured) but we can see the outcome of the throws (i.e. the macrostate consisting of the biomass distribution between the species). Nature's dice are loaded because the traits possessed by different species, in interaction with the abiotic conditions of the site, bias the outcome of each throw such that some species have a higher probability of capturing atoms than others. Dr. Wolf's die was biased because the traits of each side (the number of holes, the height, the existence of a chip) interacted with the environment (i.e. the Earth's gravitational field, the table and the surrounding air currents) to preferentially select certain outcomes more often than others and this preferential selection resulted in the average values of these traits, calculated over all throws, being different than those expected given a fair die. Where do the biases in Nature's dice come from, and how are they manifest? The answer, like so many others in ecology, was provided by Charles Darwin. This is the topic of the next chapter.

Box 4.3. **Generalized Iterative Scaling**

Generalized Iterative Scaling Algorithm and an *R* implementation: This algorithm (Della Pietra *et al.* 1997) is important because it allows one to get numerical estimates of the maximum entropy estimates without having to solve a system of simultaneous nonlinear equations; something that is difficult to do when there are more than a few constraints. The method works because, as the authors prove, the solution to the maximum (relative) entropy macrostate is the same as the solution to estimating maximum likelihood estimates for the resulting Gibbs distribution. In other words, solving the maximum likelihood problem is the same as solving the maximum entropy problem. Of course, this algorithm requires, if the states are continuous, that they be binned into discrete groups and such binning will (depending on how finely the bins are constructed) introduce some error. I also give a small function in the *R* language that implements this algorithm. A compiled Fortran program, which is much quicker for large problems (many states and many traits) is available from my website.

Input:

(1) A constraint matrix holding the values of each of the $i = 1, T$ traits (columns) for each if the $j = 1, S$ species (t_{ij}).

		1	2	. . .	S
Traits	1	t_{11}	t_{12}	. . .	t_{1S}
	2	t_{21}	t_{22}	. . .	t_{2S}
	. . .				
	T	t_{T1}	t_{T2}	. . .	t_{TS}

(2) A constraint means vector holding the community-aggregated trait values of each trait at that site.

$$(\bar{t}_1, \bar{t}_2, \ldots, \bar{t}_T)$$

Algorithm:

Initialize:

Set p_j = maximally uninformative prior values for each state.

Calculate a vector $C = (C_1, C_2, \ldots, C_T)$ where $C_i = \sum_{j=1}^{S} t_{ij}$; i.e. each C_i is the sum of the values of trait i over all the states.

Repeat for each iteration k until convergence:

For each trait (i.e. row of the constraint matrix) calculate:

$$\sum_{j=1}^{S} p_j^{(k+1)} t_{ij} = \bar{t}_i;$$

$$\gamma_i^{(k)} = \ln\left(\frac{\bar{t}_i}{\sum_{j=1}^{S}\left(p_j^{(k)} t_{ij}\right)}\right)\left(\frac{1}{C_1}\right) = \ln(1)\left(\frac{1}{C_i}\right) = 0.$$

When convergence is achieved, i.e. when the calculated probabilities are the same from one iteration to the next to within some small error,[36] then the resulting probabilities ($\hat{p}_j$) are those that simultaneously maximize the entropy conditional on the community-aggregated traits and that maximize the likelihood of the Gibbs distribution:

---

[36] For instance, you can keep iterating until $\max\left(p_i^{k+1} - p_i^k\right) < \delta$ for some very small value of δ like $0.000\,001$.

$$\hat{p}_j = \left(\frac{1}{\sum\limits_{j=1}^{S} e^{-\sum\limits_{i=1}^{T} \lambda_i t_{ij}}} \right) e^{-\sum\limits_{i=1}^{T} \lambda_i t_{ij}} = \frac{e^{-\sum\limits_{i=1}^{T} \lambda_i t_{ij}}}{Z}.$$

This means that one can solve for the Lagrange multipliers (i.e. weights on the traits, λ_i) by solving the linear system of equations:

$$\ln(\hat{p}_1) = \sum_{i=1}^{T} \lambda_i t_{i1} - \ln(Z)$$

$$\ln(\hat{p}_2) = \sum_{i=1}^{T} \lambda_i t_{i2} - \ln(Z)$$

$$\vdots$$

$$\ln(\hat{p}_s) = \sum_{i=1}^{T} \lambda_i t_{is} - \ln(Z).$$

This system of linear equations has $T + 1$ unknowns (the T values of λ plus $\ln(Z)$) and S equations. So long as the number of traits is less than $S - 1$, this system is soluble. In fact, the solution is the well-known least squares regression: simply regress the S values of $\ln(\hat{p}_j)$ on the trait values of each species in a multiple regression. The intercept is the value of $\ln(Z)$ and the slopes are the values of the Lagrange multipliers (λ_i). These slopes (Lagrange multipliers) measure by how much the ln(probability), i.e. the ln(relative abundance) changes as the value of the trait changes.

MaxEnt program in *R*

Finds p that maximizes the relative entropy subject to K equality constraints of the form $k_i = \sum_j t_{ij} p_j$ and given the maximally uninformative prior probabilities.

Example: You know only that the mean value of a 6-sided die is 3.5 and that the "trait" value assigned to each side is the number of spots ($t_{1j} = 1, 2, 3, 4, 5, 6$). Since the states are the 6 different sides, and since these are a fixed number of unordered discrete states, the maximally uninformative prior probabilities are all 1/6.

Function: MaxEnt(constraint.means,constraint.matrix, initial.probs)

constraint.means is a vector giving the K equality constraints

constraint.matrix is a matrix giving the trait values of the j states for each equality constraint.

initial.probs is a vector giving the maximally uninformative prior probabilities of each state. The default is a uniform distribution.

In the present example there is only one equality constraint: $\sum t_{1j}p_j = 3.5$ so constraint.means $= 3.5$

constraint.matrix $= 1\ \ 2\ \ 3\ \ 4\ \ 5\ \ 6$

out<-Maxent(constraint.means=c(3.5),constraint.matrix=rbind(1:6)). We could have also added initial.probs=rep(1/6,6) as an argument to the function but this is redundant in this case since the default is to assume a uniform distribution.

> out$probabilities

[1] 0.1666667 0.1666667 0.1666667 0.1666667 0.1666667 0.1666667

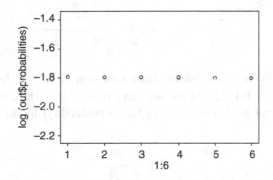

If we regress these probabilities against the trait values (i.e. 1, 2, 3, 4, 5, 6) then the slope is 0 and the intercept is -1.7918. This means that changing the values of the trait do not change the probability at all. Note that $p_j = \frac{e^{0t_i}}{e^{1.7918}} = 0.166\,66$ for all i and this is the uniform distribution: 1/6

Example 2: You know only that the mean value of a 6-sided die is 4 and that the "trait" value assigned to each side is the number of spots ($t_{1j} = 1, 2, 3, 4, 5, 6$).

out<-Maxent(constraint.means=c(4),constraint.matrix=rbind(1:6))

> out

$probabilities

[1] 0.1031964 0.1228346 0.1462100 0.1740337 0.2071522 0.2465732

$final.moments

```
        [,1]
[1,] 3.998791
    $constraint.means
    [1] 4
$constraint.matrix
     [,1] [,2] [,3] [,4] [,5] [,6]
     [1,] 1 2 3 4 5 6
```

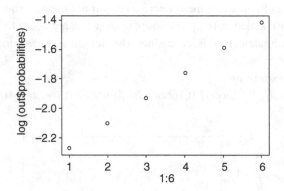

If we regress these probabilities against the trait values (i.e. 1, 2, 3, 4, 5, 6) then the slope is 0.1742 and the intercept is −2.4453. Thus, increasing the value of the trait by 1 unit increases the ln(probability) by 0.1742 units.

Note that

$$p_j = \frac{e^{0.1742 t_i}}{e^{2.4453}}$$

for all i and this is an exponential distribution.

Example 3: Now, you know only that the mean value of a 6-sided die is 4, the variance is 10 and that the "trait" value assigned to each side is the number of spots ($t_{1j} = 1, 2, 3, 4, 5, 6$).

Constraints:

$$\sum_j t_{ij} p_j = 4$$

$$\sum_j t_{2j}^2 p_j = 4$$

The second mean value constraint comes from the fact that $S^2 + \bar{x} = 10 + 4 = \sum_j x_j^2 p_j$.

```
>out<-Maxent(constraint.means=c(4,14),constraint.matrix=rbind(1:6,
(1:6)^2))
> out
$probabilities
[1] 1.180483e-10 1.188473e-04 2.223062e-01 7.725863e-01 4.988563e-03
[6] 5.984626e-08
$final.moments
[,1]
[1,] 3.782444
[2,] 14.487328
$constraint.means
[1] 4 14
$constraint.matrix
[,1] [,2] [,3] [,4] [,5] [,6]
[1,] 1 2 3 4 5 6
[2,] 1 4 9 16 25 36
```

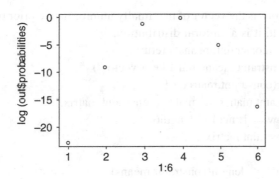

If we regress these probabilities against the two trait values (i.e. $t_1 = 1, 2, 3, 4, 5, 6$ and $t_2 = 1, 4, 9, 16, 25, 36$) then we get

$$\ln(\hat{p}) = -42.9705 + 23.2547t_1 - 3.144\ 100t_1^2$$

This function is maximal at around $t = 3.7$. Thus, increasing the value of the trait by 1 unit increases the ln(probability) by 0.1742 units.

Factoring the above equation and solving gives

$$p_j = \frac{e^{-13.6670+7.39630t_1 - t_1^2}}{1} = e^{-1(t_1 - 3.69815)^2}$$

for all i and this is the discrete version of a generalized normal distribution.

MaxEnt function

```
function(constraint.means, constraint.matrix, initial.probs =
    NA,tol=1e-8,print.lambda=F)
    {
    # MaxEnt function in R. Written by Bill Shipley
    # Bill.Shipley@USherbrooke.ca
    # Improved Iterative Scaling Algorithm from Della # Pietra, S.
    # Della Pietra, V., Lafferty, J. (1997). Inducing # features of random
    fields.
    # IEEE Transactions Pattern Analysis And Machine # Intelligence
    # 19:1-13
    #············
    # constraint.means is a vector of the community-aggregated trait values
    # constraint.matrix is a matrix with each trait defining a row and each
    # species defining a column. Thus, two traits and 6 species would produce
    # a constraint.matrix with two rows (the two traits) and 6 columns (the
    # six species)
    #initial.probs is the vector of maximally uninformative prior probabilities;
    #by default, it is a uniform distribution.
    if(!is.vector(constraint.means)) return(
            "constraint.means must be a vector")
    if(is.vector(constraint.matrix)) {
        constraint.matrix <- matrix(constraint.matrix,
            nrow = 1, ncol = length(
            constraint.matrix))
    }
    n.constraints <- length(constraint.means)
    dim.matrix <- dim(constraint.matrix)
    if(dim.matrix[2] == 1 & dim.matrix[1] > 1) {
        constraint.matrix <- t(constraint.matrix)
        dim.matrix <- dim(constraint.matrix)
    }
    if(n.constraints != dim.matrix[1])
        return("Number of constraint means not equal to number in constraint.
        matrix"
    )
    n.species <- dim.matrix[2]
    # By default, the maximally uninformative prior is a uniform distribution
    if(is.na(initial.probs)) {
```

```
        prob <- rep(1/n.species, n.species)
}
# C.values is a vector containing the sum of each trait value over the
species
C.values <- apply(constraint.matrix, 1, sum)
test1 <- test2 <- 1
while(test1 > tol & test2 > tol) {
        denom <- constraint.matrix %*% prob
        gamma.values <- log(constraint.means/denom)/
            C.values
if(sum(is.na(gamma.values)) > 0)
        return("missing values in gamma.values"
    )
        unstandardized <- exp(t(gamma.values) %*%
            constraint.matrix) * prob
        new.prob <- as.vector(unstandardized/sum(
            unstandardized))
        if(sum(is.na(new.prob)) > 0)
            return("missing values in new.prob")
        test1 <- 100 * sum((prob - new.prob)^2)
        test2 <- sum(abs(gamma.values))
        prob <- new.prob
}
lambdas<-rep(NA,n.constraints)
if(!print.lambda){
reg<-lm(log(prob)~t(constraint.matrix))}
list(probabilities = prob, final.moments = denom,
        constraint.means = constraint.means,
        constraint.matrix = constraint.matrix, entropy
        = -1 * sum(prob * log(prob)),lambdas=coef(reg))
}
```

5

Community dynamics, natural selection and the origin of community-aggregated traits

The model developed in Chapter 4, reduced to one sentence, is the following: *Find the distribution of relative abundances (i.e. theoretical probabilities) that maximizes relative entropy subject to the constraint equations.* In the previous chapter we spent a lot of time on the first half of this sentence; that is, on maximizing relative entropy. It's time to look at the second half of the sentence, namely the constraint equations. Remember that not every community-aggregated trait counts as a "constraint". A community-aggregated value is only a constraint if it biases the theoretical probabilities of gaining or losing an individual (or unit of biomass) such that some species are favored and others disadvantaged; that is, one that forces the theoretical probabilities away from those that arise in its absence.[1] A constraint equation involving i species, whose general form for equation j is $\overline{T}_j = \sum_{i=1}^{S} p_i T_{ij}$, is a function of properties (T_{ij}) of each species that bias the probability of an individual (or unit of biomass) belonging to species i. If the property (T_{ij}) is a plant trait then the average trait value, $\overline{T}_j$, is called a community-aggregated trait. A community-aggregated trait is the trait value expressed by an average entity (an individual or unit of biomass, depending on how abundance is measured) in the community. In the last chapter we largely ignored the $\overline{T}_j$ values, treating them simply as information that we were given.

It's time to peak into the black box. If we go back to the "IF – THEN" logical structure of every model, then the values of the community-aggregated traits belong to the "IF" part of our model. In order to claim that our model can explain community structure as a logical consequence of the behavior

---

[1] More specifically, an equation $\overline{T}_j = \sum p_i T_{ij}$ is a constraint if its inclusion forces **p** to differ from what it would be in its absence. As an example, consider a six-sided die. We begin with no constraints except for $\sum p_i = 1$ and this leads to $p_i = 1/6$ for all six sides (a uniform distribution). If we then specify that the mean face value, $\overline{T}$, is 3.5 then $\sum p_i T_{ij} = \overline{T}_j = 3.5$ is not a constraint because 3.5 is the predicted value of the uniform distribution and so the addition of this second "constraint" didn't change anything.

of individuals we must push this "IF" part back further by showing that community-aggregated traits, acting as active constraints, are a logical consequence of such individual behaviors. So where do community-aggregated traits come from? Why should we expect a predictable relationship between environments and such community-aggregated traits? Most importantly, since a community-aggregated trait is simply a mean (and every distribution of relative abundances has one) why should we expect this mean to constrain the relative abundances? Why should a community-aggregated trait *cause* a particular distribution of relative abundances rather than simply passively *reflect* such a distribution?

In one sense such community-aggregated traits are indeed simply "givens". As we have already seen (Chapter 3), it is an empirical fact that many of the functional traits possessed by species vary predictably along environmental gradients. Such correlations between traits and environments are the justification for much of physiological, functional and evolutionary ecology. Although only a few studies have yet explicitly measured both community-aggregated traits in different communities and the position of these communities along environmental gradients, the results of such studies have confirmed our expectation of a predictable relationship between the two. There is still an enormous amount of empirical work left to do before we can know the true strength and generality of the relationships linking environmental variables to community-aggregated traits but, in the end, such empirical relationships between community-aggregated traits and environmental gradients will probably always be the empirical "givens" of our model in any practical sense. However, if our model is to be explanatory as well as predictive, then we still need to make formal links between traits, population dynamics, community assembly and community structure. We need to demonstrate that there is some biological force that is active during community assembly and that selects for certain average traits in such a way as to bias probabilities of survival, reproduction and immigration. This is the goal of this chapter.

A game of dice

Let's go back to our dice analogy and see how far it can take us. I want to propose a simple game involving two six-sided dice that will capture some essential elements linking individual traits to population dynamics, community structure and community-aggregated traits. I will simulate this game on the computer but it is perfectly possible to play it using real dice so long as you have both fair and loaded ones.

Box 5.1. Simulation model of the dice game in the *R* language

```
function (niter=1,nspecies=6,lambda=(1:nspecies)*
0.1){
# function "dice.game"
```
#This game throws of an n-sided die, where n=nspecies and #lambda gives the bias for each side. A fair die (i.e. #neutral dynamics) would be lambda=rep(0,nspecies) mat is a #matrix of genotypes (rows) X iterations (columns) holding #the number of "individuals" per species or genotype
```
mat<-relative.abundance<-matrix(NA,nrow=6,
ncol=niter+1)
```
#initial abundances of each genotype
```
mat[,1]<-rep(60,nspecies)
relative.abundance[,1]<-rep(1/nspecies,nspecies)
```
#trait is the "trait" associated with each species or #genotype
```
trait<-1:nspecies
lnp<-(1/nspecies)-lambda*trait
p<-exp(lnp)
```
#p.birth is the (biased) probability of a "birth" for each #species or genotype
```
p.birth<-p/sum(p)
```
 # p.death is the (biased) probability of death, which is
 # 1-p.birth
```
p.death<-p.birth[nspecies:1]
```
 # add.chip and remove.chip each simulate niter random #throws of our nspecies-sided die, determining which #species or genotype will lose an individual (a death) and #which will gain an individual (a birth)
```
add.chip<- rmultinom(n=niter,size=1,prob=p.birth)
remove.chip<-rmultinom(n=niter,size=1,prob=p.
death)
```
 # now simulate the changes in "population" sizes over time
```
for(t in 1:niter){
mat[,1+t]<-mat[,t]+add.chip[,t]-remove.chip[,t]
```
#This next line resets a species with a negative number of #individuals back to zero; it could be modified (with a bit #more complexity) in order to keep the total population #size constant.
```
mat[mat[,1+t]<0,1+t]<-0
relative.abundance[,1+t]<-mat[,1+t]/sum
(mat[,1+t])
```

```
}
total.abundance<-apply(mat,2,sum)
par(mfrow=c(1,3))
#Plot the temporal dynamics of each species
plot(1:(niter+1),relative.abundance[1,],ylim=c
(0,1),xlab="Time",
ylab="Relative abundance",
type="l",lty=1,lwd=2,main="A")
for(i in 2:nspecies){
lines(1:(niter+1),relative.abundance[i,],lty=i,
lwd=2)
}
#Plot the changes in the average (community-aggregated) #trait over
time
ca.trait<-trait%*%relative.abundance
plot(1:(niter+1),ca.trait,ylim=c(0,nspecies),
xlab="Time",
ylab="Average trait value",main="B")
#Calculate the MaxEnt predicted relative abundances at the #last
iteration.
x<-Maxent(constraint.means=ca.trait[niter+1],
constraint.matrix=rbind(1:nspecies))
#Plot the observed and predicted relative abundances at the last
iteration
plot(relative.abundance[,niter+1],
x$probabilities,pch=16,
xlab="Observed relative abundance",ylab="Predicted
relative abundance",
xlim=c(0,1),ylim=c(0,1),main="C")
abline(0,1)
par(mfrow=c(1,1))
list(add.chip=add.chip,remove.chip=remove.chip,
mat=mat,
ca.trait=ca.trait,prob.birth=p.birth,prob.
death=p.death,
total.abundance=total.abundance,
relative.abundance=relative.abundance)
}
```

The game involves six players, each player being assigned one face of the die and representing a different "genotype". To start, each player is given n_0 poker chips which can represent resource units or individuals. In the simulations to follow each player starts with 60 chips. Now roll both dice. The first die simulates a death. The player whose die face is hidden (i.e. the side facing the table) must give up one poker chip if he has any left, which is placed in a common pot. This reduces the population size of his species by one individual. The second die simulates a birth. The player whose die face is on top (i.e. the face pointing up) wins the poker chip in the pot and so adds one individual to his species. By repeating these rules many times we are simulating the dynamics of the population of "individuals" in different "genotypes". Since the total number of poker chips (individuals) remains constant, the relative abundance of individuals per genotype simulates changes in community structure in a community whose total abundance is fixed as assumed, for instance, in "neutral" models. Since our die doesn't change from throw to throw our genotypes are composed of asexual organisms (i.e. no exchange of genes and no evolution) with a "heritability" of 1, i.e. perfect trait (phenotypic) transmission from parents to offspring. Box 5.1 gives a simple simulation program in the *R* language for those who want to explore the game more completely. The components of the game are shown in Figure 5.1. Let's look at each separately.

The i "traits" in Figure 5.1 are properties of each of the $j = 6$ faces of the die such as those of Dr. Wolf's die in the last chapter. The "neutral

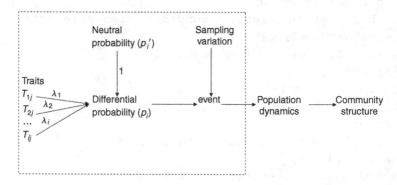

Figure 5.1. Schematic description of a simulation model linking plant traits of different species (T_{ij}) to population dynamics and community structure.

probabilities" in Figure 5.1 are the maximally uninformative prior probabilities, i.e. those that exist in the absence of any physical causes that can
bias the chances of a die landing on a given side; in this case we know that
$p_i' = 1/6$ and therefore that every genotype has exactly the same fitness
(probabilities of survival and reproduction) unless these prior probabilities
are subsequently modified. What are the λ values and the "differential
probabilities" in Figure 5.1? If you go back to the solution of the maximum
entropy problem, which is a general exponential distribution, you will recall
that

$$p_i = \frac{p_i' e^{-\sum_i \lambda_i T_{ij}}}{Z}$$

$$\ln p_i = \ln p_i' + \sum_j \lambda_j T_{ij} - \ln Z$$

$$\frac{1}{p_i} \frac{\partial p_i}{\partial T_j} = -\lambda_j. \qquad\qquad (Eqn.\ 5.1)$$

The "differential probability", or the posterior probability as a Bayesian
statistician would call it, is the probability (p_i) that results from combining
the prior probability and the changes in this prior probability that are
caused by the different properties of each face (T_{ij}). Each lambda value (λ_j)
quantifies by how much the proportional relative abundance will change
given a unit change in the trait value, when holding constant all other traits.
Thus the lambda values quantify the amount of bias in the differential
probabilities caused by the trait. If all of the lambda values are zero then the
differential probability equals the prior probability and we have a perfectly
fair die and equal fitness between genotypes. This results in "neutral"
population dynamics since each species will have exactly the same probabilities of survival and reproduction. If at least one of the lambda values is
different from zero then we have a biased die, different fitness values
between genotypes, and non-neutral population dynamics. Since we are
talking about probabilities we are not referring to a deterministic event. The
actual outcome of a given throw of the die (the "event" in Figure 5.1) will
be determined both by the probabilities of each face and by all of the
complicated, contingent, non-repeatable and unpredictable causes that vary
from throw to throw and that are described by random "sampling
variation".

A simple neutral model

I have called the maximally uninformative prior probabilities "neutral" for a reason. If the probabilities of birth and death in our game are equal for all players (1/6) then the resulting dynamics are called "neutral" in the neutral models of Hubbell (2001) and Bell (2000). So, using perfectly fair dice, our game simulates a neutral model of community assembly involving six genotypes in which random sampling variation is the only cause of changes in population dynamics and of the resulting community structure. In population genetics this is a model of genetic drift. If at least one of our lambda values is not zero then we have a non-neutral model of community assembly. What happens in our game when using either neutral or biased dice if we are dealing with a single trait (T_j): the number of spots on each face?

Figure 5.2a shows the dynamics of each population of our six genotypes when the dice are perfectly fair ($\lambda_1 = 0$) and Figure 5.2b shows the dynamics of the community-aggregated trait (the number of spots per die) using these fair dice. Using the community-aggregated trait at the end of the game, can our MaxEnt model predict the observed relative abundances? Since the community-aggregated trait value at the end of the simulation is around 3.4, we maximize the relative entropy subject to this one constraint. Figure 5.2c shows the result. As expected, the MaxEnt model has no predictive ability[2] since the trait (i.e. the number of spots per face) does not affect the probability of birth or death.

Why? Since there is no connection[3] in this simulation between the trait (i.e. the number of spots) possessed by each genotype (i.e. the face on the die) and the probability that an individual having this genotype will die or reproduce, then knowing the average trait value gives us no information about these probabilities and thus about the resulting relative abundances. Even though each genotype has exactly the same probability of an individual dying or giving birth (and therefore has the same evolutionary fitness), some genotypes increase in abundance, and other genotypes decrease in abundance, over time. This occurs because of random drift since a genotype that, by chance, gains extra individuals will tend to increase over time. To understand the link (or rather, the lack thereof) between the trait and the relative

---

[2] A perfectly uniform distribution would have a community-aggregated trait value of 3.5. Since the observed value is close to 3.5, the predicted relative abundances are all close to 1/6.

[3] In this example the connection is presumably not causal since it is unlikely that the number of spots is a physical cause of the bias in the die; rather, the number of spots is correlated with the physical imperfection as described in Chapter 4 with Dr. Wolf's die.

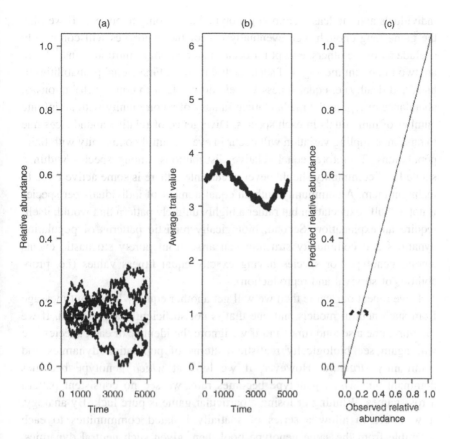

Figure 5.2. A simple neutral model of population dynamics involving six geno-types, simulated using a game with fair dice. (a) Changes in relative abundance of each genotype over 5000 generations. (b) Variation in the single community-aggregated trait ("number of spots on a face"). (c) The observed and predicted relative abundances using the maximum entropy model at generation 5000, given the community-aggregated trait. Perfect predictive ability of relative abundances would follow the 1:1 line.

abundance, simply look at Figure 5.1 and erase the arrows going from the traits and the differential probabilities. Figure 5.1 is a path diagram showing the causal links between the variables. Since, in the neutral version of our game, traits are not causally connected to the "event" and thus to population dynamics or community structure, the only cause of the outcome of an event (dying or reproducing) that differs between genotypes is sampling variation.

The resulting community structure (Figure 5.2a) shows differences in relative abundance with the dominant genotype having almost 35% of the individuals and the least common genotype having only about 5%. If we play the game long enough then eventually one of the genotypes will completely exclude all of the others, except for occasional random re-introductions.[4] There are two important messages of such neutral models. First, equal probabilities of birth and death (i.e. equal fitness values) do not lead to equal distributions of abundance except in the mathematical fantasy of a community with an infinite number of individuals in each species. Divergence of relative abundances due to random sampling variation will occur in every neutral community with finite populations. Therefore, equal relative distributions among species within a single local community should never occur unless there is some active force to maintain them. A community with an equal number of individuals per species is not a "null" expectation but rather a highly unlikely pattern that would, itself, require an explanation. Second, biologically realistic patterns of population dynamics and community structure can arise from purely stochastic events among genotypes or species having exactly equal fitness values (i.e. probabilities of survival and reproduction).

If we repeat our game then we will get another equally important message from such neutral models and one that is not sufficiently appreciated. If we run our game a second time, and if we ignore the identity of each species, we will again see biologically realistic patterns of population dynamics and community structure. However, if we look at *which* genotype becomes dominant and *which* genotype becomes rare, we see no consistent pattern across games. Winning or losing a particular game is pure luck. By analogy, if we were to allow a series of spatially isolated communities to each assemble from the same genotype pool then, given such neutral dynamics, different communities would result each time and which genotype (or species) becomes dominant or rare is completely unpredictable. If such spatially isolated communities are ordinated along an environmental variable then no systematic pattern will emerge. There will be no relationship between the typical traits of species in a community and the position of this community on the environmental gradient. In more sophisticated neutral models, in which the relative abundances of species or genotypes in the species pool (the "meta-community") are strongly uneven,[5] there will be a certain degree of correlation across communities simply because those species that are most

---

[4] This game, like neutral models, is a Markov chain. In terms of dynamic topology there are six sinks, consisting of the six states and into which any nearby dynamic trajectory will be drawn and held, and one source, consisting of equal relative abundances of the six states, and from which any nearby trajectory will be pushed away.

[5] Hubbell's model uses a logseries distribution in the metacommunity; this is explored in more detail in Chapter 7.

abundant in the species pool will usually start out at the beginning of the game with many more individuals than others. This is the basis for our "neutral" prior in Chapter 8.

However, even the more sophisticated neutral models predict two types of patterns that contradict empirical evidence. First, the degree to which communities, derived from the same regional species pool, will resemble each other will not change when comparing communities occurring in different environments. Two communities assembled at exactly the same point on an environmental gradient from the same species pool will be no more similar to each other than two communities assembled at opposite ends of the gradient;[6] see Bell *et al.* (2006) for an empirical study. This is because all dynamics are due purely to sampling fluctuations plus the metacommunity structure. Even more importantly for our purpose, and even less often appreciated, is the fact that neutral communities predict no systematic change in community-aggregated traits during assembly.[7] This is because the traits possessed by species are independent of the probabilities of survival and reproduction, which are the same for all species, and therefore of changes in relative abundance.[8] We can see this in Figure 5.2b in which the community-aggregated trait (the average number of spots) fluctuates randomly around 3.5, which is the expected value given the equal (neutral) probabilities.[9] Of course, in such an unrealistically small pool (only six genotypes) it is possible to observe a chance correlation between the trait value of each genotype and the random changes in population sizes of the genotypes over time but, with more realistic sizes of species pools such chance correlations will quickly vanish.

A non-neutral model

Now let's see what happens if the trait which is associated with each face (i.e. the number of spots) does change the probabilities of birth and death (or else is

---

[6] This assumes that there is no strong spatial autocorrelation of environmental values. Since dispersal is limited, spatially adjacent areas of space would have some similarity in composition in more sophisticated neutral models because of similarity in the initial relative abundances.

[7] If genotypes that are already locally extinct due to random drift cannot subsequently reinvade then eventually only one genotype will exist and so the community-aggregated trait will eventually begin to converge on the trait value of this genotype.

[8] This can be made more precise. A diagram like Figure 5.1 is called a "directed acyclic graph". Two variables in such a graph are independent if they are "*d*-separated". If the lambda values are zero then the traits are *d*-separated from the probabilities, population dynamics and community structure. See Chapter 2 of Shipley (Shipley 2000a) for more details.

[9] $\bar{T} = \sum p_i T_i = \frac{1}{6} \sum_{i=1}^{6} i = 3.5.$

Table 5.1. *Values in columns 3–6 only hold until at least one genotype is removed.*

Genotype	Birth probability (b)	Death probability (d)	Probability of increasing by 1	Probability of remaining constant	Probability of decreasing by 1
1	0.39	0.01	0.386	0.608	0.006
2	0.29	0.04	0.278	0.693	0.028
3	0.18	0.09	0.164	0.762	0.074
4	0.09	0.18	0.074	0.762	0.164
5	0.04	0.29	0.028	0.693	0.278
6	0.01	0.39	0.006	0.608	0.386

correlated with some trait that is causally connected with these probabilities). Remember that the face of the die pointing down onto the table determines the genotype that will lose an individual (assuming that it still has some) while the face landing up on the second die determines the genotype that will gain an individual. By defining T_{i1} as the number of spots on each face, and choosing $\lambda = -0.1$ in Equation 5.1, we get probabilities of birth and death (and thus fitness) for each genotype that are shown in Table 5.1.

Remembering that the opposite sides of a six-sided die always sum to 7, this means that the probability of genotype 1 experiencing the "death" of an individual[10] (i.e. face 1 landing down on the table) is one minus its probability of giving "birth" to a new individual and so on. Since fitness is the relative number of expected offspring per individual, genotype 1 has the highest fitness while genotype 6 has the lowest fitness. Of course, once some genotypes have been completely removed from the game, the probabilities of death for those genotypes that are still present will increase. In our simple game the eventual outcome, like for the neutral model, is that one genotype will eventually completely dominate the community except for rare re-introductions by the others. There are two differences however. First, the change in composition occurs much more rapidly. Second, the winning genotype will almost always be genotype 1 in every repetition of the game. Figure 5.3 shows the resulting outcome of this game.

The behavior of this non-neutral model is quite different. First, although there is still sampling variation, the outcome is now quite predictable: genotype 1 dominates, genotypes 2 and 3 (each of which have greater probabilities of increasing than of decreasing) initially increase while genotypes 4, 5 and 6 decrease. As the less fit genotypes are excluded, genotypes 3 and then later 2

---

[10] While it has at least one individual.

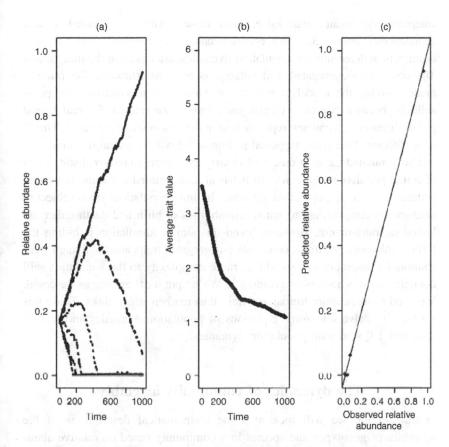

Figure 5.3. A simple non-neutral model of population dynamics involving six genotypes, simulated using a game with loaded dice. Probabilities are given in Table 5.1. (a) Changes in relative abundance of each genotype over 5000 generations. (b) Variation in the single community-aggregated trait ("number of spots on a face"). (c) The observed and predicted relative abundances using the maximum entropy model at generation 5000, given the community-aggregated trait. Perfect predictive ability of relative abundances would follow the 1:1 line.

begin to decrease. Eventually the community is dominated by genotype 1 except for random and rare re-introductions that are quickly extinguished. This pattern is repeatable from game to game. More importantly, we see a predictable change in the community-aggregated trait over time. It decreases as the genotypes having lower trait values increase in relative abundance, slowing down as the changes in relative abundance decrease, and essentially stops changing when the community is overwhelmingly dominated by genotype 1. The predicted relative abundances of the MaxEnt model, given the final community-aggregated trait value, agree closely with the observed relative abundances (Figure 5.3c). This occurs because the probabilities of birth and death, which determine the population dynamics, are linked to the trait values. The community-aggregated trait value acts as a *constraint* in the MaxEnt model, forcing the model to bias the maximally uninformative prior probabilities, because community-aggregated trait value reflects the real biased probabilities of survival and reproduction in the dynamics of the game. This is quite different than what happened in Figure 5.2 with a neutral dynamic.

Our simulated game of dice is obviously too simple to be a realistic model of actual population dynamics but it has at least pointed out some important patterns concerning population dynamics in finite populations (thus subject to random sampling variation) when probabilities of birth and death either are linked to traits or not. However, even this simple simulation is hiding the formal links between traits, community-aggregated traits and the dynamics of community assembly and so adding more complexity to the simulation will not help. There's no way of avoiding it. We've put it off as long as we could. We need to forge more formal, explicit, thus mathematical, links. To do this we have to delve into some equations of population dynamics and natural selection. I'll start with population dynamics.

The dynamics of community assembly

In this section we will look at some mathematical descriptions of the dynamics of genotypes and species in a community based on relative abundances. Any mathematical description of population dynamics must start with the per capita growth rate of a population, which is the instantaneous average number of new individuals added to (or removed from) the population by each existing individual at time t. The definition for the per capita rate of population change of genotype i at time t is $r_i(t) = \frac{1}{n_i(t)}\frac{dn_i(t)}{dt}$. Note that $r_i(t)$ can change over time and so this equation is a relationship that is true by definition and requires no empirical support. Also by definition, the relative abundance of genotype i at time t is $p_i(t) = \frac{n_i(t)}{N(t)}$, where $N(t)$ is the total number of individuals of all genotypes at time t and $n_i(t)$ is the number of individuals having genotype i at time t. Combining these two definitions, it follows (Shipley 1987a) that the rate of change of relative abundances (Box 5.2) can be expressed as $\frac{d\ln(p_i(t))}{dt} = (r_i(t) - \bar{r}(t))$ or, equivalently, as:

$$\frac{dp_i(t)}{dt} = p_i(t)\cdot(r_i(t) - \bar{r}(t)). \qquad \text{(Eqn. 5.2)}$$

Equation 5.2 contains no biological assumptions and so it, too, is true by definition.

Box 5.2. **Derivation of Equation 5.2**

Substituting the definition or relative abundance into the definition of instantaneous per capita growth rate we get:

$$r_i(t) = \frac{1}{n_i(t)}\frac{dn_i(t)}{dt} = \frac{1}{p_i(t)N(t)}\frac{d(p_i(t)N(t))}{dt}$$

$$r_i(t) = \frac{1}{p_i(t)N(t)}\left[N(t)\frac{dp_i(t)}{dt} + p_i(t)\frac{dN(t)}{dt}\right] = \frac{1}{p_i(t)}\frac{dp_i(t)}{dt} + \frac{1}{N(t)}\frac{dN(t)}{dt}.$$

Also by definition, the community-aggregated per capita growth rate is $\overline{r}(t) = \sum p_i(t)r_i(t)$. Now, since

$$\overline{r}(t) = \sum p_i(t)r_i(t) = \sum p_i(t)\frac{1}{n_i(t)}\frac{dn_i(t)}{dt}$$

$$= \sum p_i(t)\frac{1}{p_i(t)N(t)}\frac{d(p_i(t)N(t))}{dt},$$

it follows that:

$$\overline{r}(t) = \sum \frac{1}{N(t)}\left[N(t)\frac{dp_i(t)}{dt} + p_i(t)\frac{dN(t)}{dt}\right] = \sum\left[\frac{dp_i(t)}{dt} + \frac{p_i(t)}{N(t)}\frac{dN(t)}{dt}\right].$$

However, $\sum \frac{dp_i(t)}{dt} = 0$ since the proportions always sum to unity and so the sum of the changes in proportions must sum to zero. Also, since

$$\sum p_i(t)\frac{1}{N(t)}\frac{dN(t)}{dt} = \frac{1}{N(t)}\frac{dN(t)}{dt}\sum p_i(t) = \frac{1}{N(t)}\frac{dN(t)}{dt},$$

it follows that $r_i(t) = \frac{1}{p_i(t)}\frac{dp_i(t)}{dt} + \overline{r}(t)$, from which we get

$$\frac{dp_i(t)}{dt} = p_i(t)\left(r_i(t) - \overline{r}(t)\right)$$

Equation 5.2 is the fundamental equation of dynamic game theory and is sometimes called the "replicator" equation of natural selection (Nowak 2006). It was first introduced into population dynamics by van Hulst (1992). The resulting system of equations, when we write down an equation for each genotype in the community, will express all the complicated dynamic behavior, including deterministic chaos, which we saw in Chapter 2. In fact, although not immediately obvious, the classic Lotka–Volterra equations of community dynamics as well as Tilman's models (Chapter 2) are a special cases of these replicator equations (Shipley 1987a).

In this replicator equation we find a term $(r_i(t) - \overline{r(t)})$ expressing the difference between the per capita growth rate of species i and the community-aggregated per capita growth rate. This difference is often equated with relative fitness of different genotypes and we can see why: genotypes whose per capita growth rate is greater than the average at time t will increase in relative abundance (they will be selected for) while genotypes whose per capita growth rate is less than the average at time t will decrease in relative abundance (they will be selected against). However, it is wrong to assume an equivalence between fitness and this difference since we know (Figure 5.2a) that genotypes can have different realized per capita growth rates even when the probabilities of birth and death are equal (i.e. when fitness is equal). The difference $r_i(t) - \overline{r(t)}$ is more properly described as an *estimate* of fitness based on realized per capita birth and death schedules. The realized per capita birth and death schedules include a random component attributable to sampling fluctuations while the fitness is the theoretical per capita growth rate that is determined by the probabilities of birth and death.

How do we link plant traits to per capita growth rates? In Chapter 2 we saw how species-based models of community assembly started with population dynamics expressed as purely phenomenological equations (for instance, the Lotka–Volterra equations) and became more and more mechanistic. This eventually led to Tilman's model (1988) in which per capita growth rates are expressed as a function of plant traits. In order to link per capita growth rates to traits we can therefore express the $r_i(t)$ values as functions of the phenotypic traits, which are the same as the genotype traits in our game: $r_i(t) = f(\mathbf{T}_i) + \varepsilon_i(t)$; $\mathbf{T}_i$ is the vector of trait values for genotype i.

For mathematical simplicity we will assume a linear and non-interacting function of trait values at time t so that we can regress the $r_i(t)$ values of the different genotypes at time t on the j trait values of each genotype (T_{ij}). For more complicated functions linking traits to per capita growth rates we would have to use generalized additive models (Hasti and Tibshirani 1990, Yee and Mitchell 1991). Such complicated functions do not change the basic logic. Doing this we get, for a linear model: $r_i(t) = f(\mathbf{T}_i) + \varepsilon_i(t) = a(t) + \sum_k \beta_k(t)T_{ik} + \varepsilon_i(t)$, where the $\beta_k(t)$ values are the partial regression slopes of each of the k traits on the per capita growth rate at time t and $\varepsilon_i(t)$ is the deviation of the per capita rate of population change of genotype i from the expected value at time t due to random sampling variation. The predicted value of the per capita growth rate of genotype i given the traits, $\hat{r}_i(t) = a(t) + \sum_k \beta_k(t)T_{ik}$, is then the absolute fitness of genotype i at time t expressed as a function of all of those traits biasing probabilities of birth and death; this is because the random component of $r_i(t)$ that is due to sampling fluctuations has been removed. Using our expression for $r_i(t)$ and substituting it into our replicator equation of natural selection, we get (Box 5.2):

$$\frac{d \ln p_i(t)}{dt} = \frac{1}{p_i(t)}\frac{dp_i(t)}{dt} = \sum_k \beta_k(t)\left(T_{ik} - \overline{T}_k(t)\right) + \varepsilon_i(t). \qquad \text{(Eqn. 5.3)}$$

This is our first formal link between the population dynamics of different genotypes, changes in community structure (i.e. the relative abundance of the genotypes), the traits possessed by each genotype, the community-aggregated traits, and the effect of random sampling effects. It says that the proportional change in the relative abundance of a genotype relative to a given ensemble of genotypes at any time t is determined by three things: (i) the effect, $\beta_k(t)$, of each trait in determining the per capita growth rate of the genotype at that time, (ii) the difference in its trait values relative to the average trait values of the group (a **community-aggregated trait** value), and (iii) the random effects of sampling from a finite population ($\varepsilon_i(t)$).

Intraspecific versus interspecific trait values

Some readers might have noticed a mathematical slight of hand in this last section. We started out taking about "species" and suddenly switched to "genotypes". More skeptical readers (not *you* . . .) might suspect that this was the prestidigitation of a con artist so let me assure you that it is also possible to express our relationship as average values of genotypes between each species.

We first group together all those genotypes in the community belonging to species i and calculate the average per capita growth rate of these genotypes ($\overline{r_i(t)}$). We then calculate the average trait values of each of the k traits of those j genotypes (T_{ijk}) in species i ($\overline{T_{ik}}$). Doing so yields the relevant equations at the species level (Box 5.3):

$$\frac{1}{p_i(t)}\frac{dp_i(t)}{dt} = \frac{d \ln(p_i(t))}{dt} = \overline{r_i(t)} - \overline{r(t)}$$

$$\frac{1}{p_{ij}(t)}\frac{dp_{ij}(t)}{dt} = \frac{d \ln(p_{ij}(t))}{dt} = \sum_k \beta_k \left(\left(T_{ijk} - \overline{T_{ik}}\right) + \left(\overline{T_{ik}} - \overline{T}_k\right)\right) + \varepsilon_{ij}.$$

$$\text{(Eqn. 5.4a, b)}$$

Equation 5.4a says that the proportional change in the relative abundances at the species level is given by the difference in the per capita growth rate of the average genotype in that species minus the per capita growth rate of the average genotype in the community.[11] The community-aggregated per capita growth rate will normally be close to zero if the total number of individuals is close to the carrying capacity of the environment and will be exactly zero if the total number of individuals is constant over time. Equation 5.4b says that the proportional change in the relative abundance of genotype j of species i is a weighted sum of two differences plus sampling variation. The first difference is that of the genotype trait relative to the mean trait of its species (intraspecific selection). The second difference is that of its species mean trait relative to the community-aggregated trait (interspecific selection). If the differences in trait means between species are much greater than the differences in traits between genotypes of the same species, then we can pretty much ignore the intraspecific differences and community dynamics will be primarily determined by differences in interspecific trait means. In fact, as we will see later when we get to the notion of heritability, this is especially true since the population genetic measure of heritability will generally be higher when comparing genotypes across species. The degree to which we can cheat by using species means will depend on the proportion of the total variation in a trait that is due to intraspecific versus interspecific sources and this can be quantified by variance components.

Box 5.3. **Derivation of Equations 5.3 and 5.4**

Given (i) $\frac{d\ln(p_i(t))}{dt} = \left(r_i(t) - \overline{r(t)} \right)$.

Given (ii) $r_i(t) = a + \sum_j \beta_j(t) T_{ij} + \varepsilon_i$.

Given (iii) $\overline{\varepsilon} = 0$.

Given (iv) $\overline{r(t)} = \sum_i p_i(t) \left(a + \sum_j \beta_j(t) T_{ij} + \varepsilon_i \right) = a + \sum_j \beta_j(t) \overline{T}_{ij}$.

It follows that

$$\frac{d\ln p_i(t)}{dt} = \frac{1}{p_i(t)} \frac{dp_i(t)}{dt} = \left(a + \sum_j \beta_j(t) T_{ij} + \varepsilon_i \right) - \left(a + \sum_j \beta_j(t) \overline{T}_j \right)$$

$$\frac{d\ln p_i(t)}{dt} = \sum_j \beta_j(t) \left(T_{ij} - \overline{T}_j \right) + \varepsilon_i,$$

which is Equation 5.3.

---

[11] Here I have to raise a red flag. Evolutionary ecologists, reading this section, will immediately object that natural selection only acts on genotypes of the same species, not on genotypes of different species. This is wrong, as I will explain in the next section. It is true that *evolution* by natural selection only acts on genotypes of the same species, and this is why evolutionary ecologists correctly look only at intraspecific heritable variation, but evolution is only one consequence of natural selection.

Now, to get average values per species we group together the j genotypes of each of the i species and note that

$$\frac{1}{n_i(t)}\frac{dn_i(t)}{dt} = \left(\frac{1}{\sum_j n_{ij}(t)}\right)\left(\frac{d\left(\sum_j n_{ij}(t)\right)}{dt}\right) = \left(\frac{1}{\sum_j n_{ij}(t)}\right)\left(\sum_j \frac{dn_{ij}(t)}{dt}\right)$$

$$= \left(\frac{1}{\sum_j n_{ij}(t)}\right)\left(n_{i1}(t)\left(\frac{1}{n_{i1}(t)}\right)\left(\frac{dn_{i1}(t)}{dt}\right) + \cdots + n_{ij}(t)\left(\frac{1}{n_{ij}(t)}\right)\left(\frac{dn_{ij}(t)}{dt}\right)\right)$$

$$= \left(\frac{1}{\sum_j n_{ij}(t)}\right)\left(n_{i1}(t)r_{i1}(t) + \cdots + n_{ij}(t)r_{ij}(t)\right)$$

$$= \sum_j p_{ij}(t)r_{ij}(t)$$

$$= \overline{r_i(t)}.$$

Now, to get the equation linking the change in relative abundances at the level of species with the intraspecific and interspecific traits we begin with

$$\frac{1}{p_{ij}(t)}\frac{dp_{ij}(t)}{dt} = r_{ij}(t) - \overline{r(t)} - \left(r_{ij}(t) - \overline{r_i(t)}\right) - \left(\overline{r_i(t)} - \overline{r(t)}\right).$$

Since $r_{ij}(t) = a + \sum_k \beta_k(t)T_{ijk}(t) + \varepsilon_{ij}$,

since $\overline{r_i(t)} = a + \sum_k \beta_k \overline{T_{ik}}$

and since $\overline{r(t)} = a + \sum_k \beta_k \overline{T_k}$, therefore

$\frac{1}{p_{ij}(t)}\frac{dp_{ij}(t)}{dt} = \sum_k \beta_k\left[\left(T_{ijk} - \overline{T_{ik}}\right) + \left(\overline{T_{ik}} - \overline{T_k}\right)\right] + \varepsilon_{ij}$, which is Equation 5.4.

Equations 5.3 and 5.4 describe *how* population dynamics depend on species traits and community-aggregated traits but not *why* the link is biologically generated. The *why*, which will be developed next, is simply stated: natural selection. Most people equate "natural selection" with evolution but this is wrong. Here is how Darwin defined it (Darwin 1859): "... *can we doubt (remembering that many more individuals are born that can possibly survive) that individuals having any advantage, however slight, over others, would have the best chance of surviving and of procreating their kind? On the other hand, we may feel sure that any variation in the least degree injurious would be rigidly destroyed. This preservation of favourable individual differences and variations, and the destruction of those which are injurious, I have called Natural Selection."* Natural selection is simply the consequence of differential probabilities of survival and reproduction across genotypes when these probabilities are functions of organismal traits. It is the outcome of Nature playing with loaded dice and leads to two consequences.

The first consequence, when comparing genotypes within the same interbreeding group (i.e. species), is the evolution of new phenotypes and, when reproductive barriers form, the evolution of new species. This is, of course, the most studied aspect of natural selection and is why "natural selection" and "evolution" are so tightly linked in biology. What does natural selection have to do with community assembly? Certainly we community ecologists recognize that the species we list in our community descriptions, and the morphological, physiological and phenological attributes that each species possesses, arise through evolution. Certainly we invoke natural selection to explain the *origin* of niche partitioning between species. However, we do not generally invoke natural selection to explain and predict the *process* of community assembly at a site over ecological time scales or the changes in relative abundance of different species during succession or along environmental gradients. Natural selection, community ecologists might be tempted to say, is responsible for setting the ecological stage but not for directing the play. Presumably this is because such community ecological processes occur on much shorter time scales than are generally needed for the adaptive radiation of different species. If so, then this is only because we mentally – and incorrectly – equate natural selection with the evolution of new species. Natural selection, as Darwin described it, is simply the consequence of differential probabilities of survival and reproduction across genotypes when these probabilities are functions of organismal traits. Genotypes can be grouped into those that can exchange genes (genotypes belonging to the same species) and those that cannot (genotypes belonging to different species). Natural selection occurring between genotypes of the same species *causes* evolution but natural selection is not a *synonym* of evolution. Natural selection can have other biological consequences besides evolution.

One consequence of natural selection besides evolution, when occurring between genotypes of different species, is community dynamics. This is because, when comparing genotypes occurring in different species, the differential probabilities of survival and reproduction of these genotypes lead to changes in relative abundances between these genotypes and, since these genotypes are grouped into different species (Equation 5.3), this in turn leads to changes in relative abundances between species. However, as explained above, "differential probabilities of survival and reproduction" is simply what Darwin called "natural selection". Therefore natural selection, occurring between genotypes found at the same site but grouped into different species, causes changes in interspecific population dynamics and this causes changes in community structure. This link between natural selection and changes in community structure might seem suspect but it was actually made in the original Darwin–Wallace papers, communicated to the Linnean Society of London on June 30th, 1858 in which Charles Darwin and Alfred Wallace jointly published their theory of evolution by natural selection (Darwin and Wallace 1858). Here is how Wallace described natural selection on page 54 of that publication: *"The life of wild animals is a struggle for existence. The full exertion of all their faculties and all their energies is required to preserve their own existence and provide for that of their infant offspring. The possibility of procuring food during the least favourable seasons, and of escaping the attacks of their most dangerous enemies, are the primary conditions which determine the existence of both individuals and entire species. These conditions will also determine the population of a species; and by careful consideration of all the circumstances we may be enabled to comprehend, and in some degree to explain, what at first sight appears so inexplicable – the excessive abundance of some species, while others closely allied to them are very rare"*. There you have it. My suspect claim that natural selection among genotypes of different species leads to community structure ("the excessive abundance of some species while others are rare") is found in the very first official publication of Darwin and Wallace.

In order to further link natural selection of genotypes to community dynamics we need to look at the Breeder's Equation of quantitative genetics. We will therefore make a detour for a few pages to better understand how phenotypic selection is measured by quantitative geneticists and then return to make the link with community dynamics.

Quantifying natural selection: the Breeder's Equation

If we compare the value of some heritable trait (T) in a collection of parents and their offspring in any group of organisms then we expect to see a positive relationship between the two: children tend to resemble their own parents more than they resemble randomly chosen parents in the group. In Figure 5.4a we see the relationship between the value of some quantitative trait (a phenotypic attribute) in 26 offspring and the average value of the trait (the midparent value) of the two parents of each offspring. In Figure 5.4b I have plotted these data in a slightly different way by subtracting the

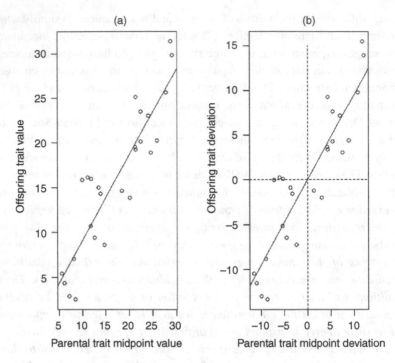

Figure 5.4. A typical parent–offspring regression for a continuous trait, shown in original units (a) and using trait values centered about their means (b).

average value of the parental midpoint values from both parents and off-spring.[12] The regression line is now given by $(\hat{T}_{oi} - \overline{T}_p) = h^2(T_{pi} - \overline{T}_p)$ where $\hat{T}_{oi}$ is the predicted (i.e. average) value of the offspring of the ith parents given the parental midpoint (T_{pi}) of these parents; $\overline{T}_p$ is the average value of the parental midpoint values, and h^2 is the slope of the relationship. The slope (h^2) is called the *heritability*[13] of the trait in quantitative genetics and

[12] Francis Galton, a cousin and contemporary of Charles Darwin, produced a graph similar to this example by plotting adult human heights of offspring and (mid-) parents. Because the slope was less than unity, Galton noted that parents who were taller than average tended to produce children who, although still taller than average, were *shorter* than their parents. Similarly, parents who were shorter than average tended to produce children who, although also shorter than average, where *taller* than their parents. Since the trait values of offspring, irrespective of those of their parents, seem to be moving towards the populational mean, he called this a "regression" towards the mean: thus the origin of the term "regression" for this statistical method. Galton further argued that this trend would erode away any selective progress thus preventing Darwinian evolution. Pearson (1903) pointed out the error in logic which we will see next.

[13] More specifically, the *narrow-sense* heritability.

varies[14] between 0 and 1. A heritability of exactly 1 means that the phenotypic value of the trait in an average offspring $(\hat{T}_{oi})$ of a pair of parents is exactly the same as the average value of its own two parents (T_{pi}), thus when $h^2 = 1$ then $\hat{T}_{oi} = 1 T_{pi}$. A heritability of exactly 0 means that the phenotypic value of the trait in an average offspring of a pair of parents is no more similar to the midpoint of its own parents than it is to any two randomly chosen parents in the population (i.e. $\overline{T}_p$); thus when $h^2 = 0$ then $\hat{T}_{oi} = \overline{T}_p$. By definition the heritability is a regression slope and so it is a statistical, not a strictly mechanistic, notion. Also, by definition, it is the ratio of the covariance of the trait T between offspring and parents, $\text{Cov}(T_o, T_p)$, and the variance of the trait in the parents, $\text{Var}(T_p)$; thus $h^2 = \text{Cov}(T_o, T_p)/\text{Var}(T_p)$. The offspring–parent covariance is also called the "additive genetic variance".

So the relationship between the phenotypic value expressed by the offspring is given by $\hat{T}_{oi} = h^2(T_{pi} - \overline{T}_p) + \overline{T}_p$. Now imagine that only some of the parents survive to reproduce and that the chance of a parent surviving increases with increasing values of T; in other words, increasing values of trait T provide a selective advantage. Figure 5.5 shows this for one such scenario. Those parents who survive and then reproduce (the dark circles in Figure 5.5) are indicated as T_p^* and their offspring are indicated as T_o^* while those parents who die (and so are removed from the population without reproducing) are shown by the open circles.

Now, in the absence of selection we have $\hat{T}_{oi} = h^2(T_{pi} - \overline{T}_p) + \overline{T}_p$ but in the presence of selection we have $\hat{T}_{oi}^* = h^2(T_{pi}^* - \overline{T}_p) + \overline{T}_p$. The change due to selection $(\hat{T}_{oi}^* - \hat{T}_{oi})$ is therefore given by $h^2(T_{pi}^* - T_{pi})$. In Figure 5.5 the average parental midpoint deviation in the subset of parents who survived was 4.65 (the thick solid cross) versus a midparent value of 0 for the parental population before selection (the thin broken cross). This difference between average parental values before and after selection, $\overline{T}_p^* - \overline{T}_p$, is called the *selection differential*[15] (S) and is shown by the arrow in Figure 5.5. The equation describing the change in the average trait value in the offspring caused by selection is the classic *Breeder's Equation*: $\Delta \overline{T} = h^2 S$. Natural selection simply means that some parents have a higher probability of surviving and/or of producing more offspring than others and that this

---

[14] Actually, the estimated value of h^2 can go beyond this range, but only because of random sampling variation or measurement error.

[15] By convention, one speaks of a selection "differential" if we are dealing with only one trait, and a selection "gradient" when dealing with multiple traits.

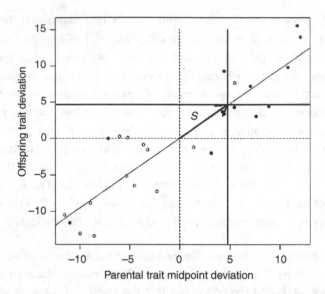

Figure 5.5. A parent–offspring regression including natural selection. The open circles represent parents who die before reproducing (i.e. are selected against) while the filled circles represent parents who survive to reproduce. The dotted cross marks the bivariate mean in the absence of selection and the solid cross marks the bivariate mean in the presence of selection. The arrow shows the vector of change in the bivariate means, and this is the selection gradient (*S*).

probability bias is associated with heritable traits. In Figure 5.5 I have simulated the case where only survival probabilities differ. If there was differential reproductive success associated with trait *T* then parents with certain values of *T* would generate more points (offspring) and this too would shift the average parental average.

Before going on, let's pause and look at the Breeder's Equation again. Since the selection event occurs at time t while the offspring are produced at time $t+1$, we can rewrite our equation as $\overline{T}(t+1) = h^2(t)\left(\overline{T}^*(t) - \overline{T}(t)\right) + \overline{T}(t) = h^2(t)S(t) + \overline{T}(t)$. If a series of selection events occur over time $(0,1, 2, \ldots t)$ then the mean trait value at time t will be

$$\overline{T}(t) = \sum_{j=0}^{t} h^2(j)S(j) + \overline{T}(0). \qquad \text{(Eqn. 5.5)}$$

Stop and think about what this simple, almost self-evident, equation is telling us. It is saying that, in the presence of differential probabilities of survival and reproduction that are associated with the trait, the average trait value $(\overline{T}(t))$ will change over time and that this change is a *consequence* of, and is therefore *constrained* by, the selective pressures and heritabilities experienced by individuals over time. $\overline{T}(0)$ is the community-aggregated trait value at the start of the process and so is the average trait value in the absence of selection. $\overline{T}(0)$ is the average trait value in a neutral scenario. Notice that each of the trait means at time t is, by definition, $\overline{T}(t) = \sum_{i=1}^{G} p_i(t) T_i$ where the summation is over all genotypes. But this is simply a community-aggregated trait!

At the community level these different genotypes will be grouped into different species. This subdivision of genotypes into reproductively isolated groups (species) is irrelevant to the process of natural selection *in general* even though it is central to the process of natural selection *as a cause of evolution*. Natural selection of variable genotypes within a species leads to evolution because only at this level can new favorable mutations be transmitted from parents to offspring. Natural selection of variable genotypes between species does not lead to evolution. This is because favorable mutations are prevented from crossing the reproductive barrier. However, natural selection still leads to differential survival and reproduction of the genotypes grouped into the various species (differential probabilities of survival and reproduction *is* natural selection) and this leads to community dynamics as Wallace noted.

If all variation in the trait value exists at the interspecific level (i.e. if all individuals within a species have exactly the same trait value) then there would be no evolution but there would still be changes in the relative abundance of species in the community. If all variation exists at the intraspecific level (i.e. the mean trait value is the same for all species) then there would be evolution but no changes in the relative abundance of species in the community.[16] In reality, both of these processes are occurring at the same time. The process of natural selection, in which differently adapted traits possessed by genotypes in a given environment result in biased probabilities of survival and reproduction, results in constraints on mean trait values at the community level, as given in the Breeder's Equation. This is the origin of our community-aggregated traits.

If fitness differences between genotypes are determined by more than one (say, m) traits then we have a vector of trait values for each genotype: $\mathbf{T} = (T_i, \dots, T_m)$. These traits are not necessarily independent of one another.

---

[16] Assuming that no new favorable mutations arise in only some species, but this eventuality is excluded by the assumption that the mean trait value is the same for all species.

In this case one obtains the multivariate version of the Breeder's Equation (Box 5.4). The multivariate version, expressed in terms of the relationship between genotypic fitness and this group of traits, is $\Delta \overline{\mathbf{T}}(\Delta t) = \mathbf{P}(t)\boldsymbol{\beta}(t)$,[17] where $\Delta \overline{\mathbf{T}}(\Delta t)$ is the vector of changes in the trait means due to selection during the time interval Δt, $\mathbf{P}(t)$ is the additive variance–covariance matrix of the phenotypic traits in the adults at time t, and $\boldsymbol{\beta}(t)$ is the vector of partial regression coefficients, representing the linear and nonlinear components of the selection gradient, that result from regressing relative fitness against the entire set of phenotypic traits at time t. The equation, as I have given it, emphasizes that neither the covariance between traits nor the relationship between traits and fitness are necessarily constant over time, although they will be approximately constant over short time intervals. In applications of the Lande and Arnold (1983) model of phenotypic selection, based on the multivariate Breeder's Equation, one generally regresses some indicator of fitness such as the lifetime reproductive output of an individual, often standardized for the population, on the vector of midparent trait values of the parents. However, an even better indicator of genotype fitness would be the per capita growth rate of the genotype relative to the per capita growth rate of all genotypes together; i.e. $r_i(t) - \overline{r(t)}$. Therefore the relationship between the vector of mean trait values before $(\overline{\mathbf{T}}(t))$ after $(\overline{\mathbf{T}}(t+1))$ selection is $\overline{\mathbf{T}}(t+1) = \mathbf{P}(t)\boldsymbol{\beta}(t) + \overline{\mathbf{T}}(t)$.

Stabilizing or disruptive selection

The above Breeder's Equation describes directional selection. That is, selection that causes a change in the mean value of the trait. However, selection can also be stabilizing or disruptive as well as, or even independently of, its effect on means. Stabilizing selection occurs when traits close to the mean have a selective advantage with respect to those farther from the mean, thus reducing the variance of the trait. Disruptive selection occurs when traits close to the mean have a selective disadvantage with respect to those farther from the mean, thus increasing the variance of the trait. How can we model changes in trait values when selection is acting on the variances and

---

[17] An equivalent, but alternative, formulation is $\Delta \overline{\mathbf{T}}(\Delta t) = \mathbf{G}(t)\mathbf{P}(t)^{-1}\mathbf{S}(t)$, where $\mathbf{G}$ is the matrix of genetic (i.e. parent–offspring) covariances between the traits, $\mathbf{P}$ is the matrix of phenotypic (i.e. parental) variances and covariances, and $\mathbf{S}$, the selection gradient, is a vector of selection differentials.

Box 5.4. **The derivation of the Breeder's Equation in terms of relative fitness**

This derivation follows from Lande and Arnold (1983). Let $W_i(X)$ be the absolute fitness of individual i having genotypic trait value X; for instance $W_i(X) = r_i(X, t)$. The average (expected) absolute fitness of genotype X is $\overline{W(X)}$ and $W_i(X) = \overline{W(X)} + e_i(X)$, where $\overline{e(X)} = 0$. In other words, individual absolute fitness is decomposed into the average absolute fitness of its genotype plus its individual deviation from this average. Let $\mathbf{p}(X)$ be the proportion of individuals having each possible genotype (X) before the selection event; this is the distribution of trait values before selection. The average (expected) fitness of all individuals over all genotypes is $\overline{W} = \sum_X p(X)\overline{W(X)}$. The relative fitness of genotype X is $\overline{w(X)} = \overline{W(X)}/\overline{W}$, where $\overline{w} = 0$. If the distribution of trait values before selection is $\mathbf{p}(X)$, then the distribution of trait values after the selection event is $\mathbf{p}^*(X) = \mathbf{w}(X)\mathbf{p}(X)$. The average trait value before selection is $\overline{X} = \sum_X Xp(X)$, the average trait value after selection is $\overline{X^*} = \sum_X Xp^*(X) = \sum_X Xw(X)p(X)$ and the average relative fitness is $\overline{w} = \sum_X w(X)p(X)$. The directional selection differential (S, i.e. the change in the average trait value in the offspring generation due to the selection) is $S = \overline{X^*} - \overline{X} = \sum_X Xw(X)p(X) - \sum_X Xp(X)$.

Now,

$\sigma_{w(X), X} = E[(w(X) - \overline{w})(X - \overline{X})]$

$\sigma_{w(X), X} = E[w(X)(X - \overline{X})]$

$\sigma_{w(X), X} = E[Xw(X)] - E[\overline{X}w(X)] = E[Xw(X)] - \overline{X}$

$\sigma_{w(X), X} = \sum_x Xw(X)p(X) - \sum_X Xp(X)$

$\sigma_{w(X), X} = S$.

So, $\Delta\overline{T}(t+1) = h^2(t)S(t) = h^2(t)\sigma_{w(X), X}$. But $h^2(t)$ is the slope of the parent–offspring trait relationship, so

$$\Delta\overline{T}(t+1) = \left(\frac{\sigma_{X_O, X_P}}{\sigma_X^2}\right)\sigma_{w(X), X} = \sigma_{X_O, X_P}\left(\frac{\sigma_{w(X), X}}{\sigma_X^2}\right) = \sigma_{X_O, X_P}\beta.$$

This is the univariate version of the Breeder's Equation but can be directly extended to the multivariate case by replacing with the appropriate matrices:

$$\Delta\overline{\mathbf{T}}(t+1) = \mathbf{P}(t)\boldsymbol{\beta}(t).$$

covariances (thus the second moments of the trait distribution) rather than on the means (thus the first moments of the trait distribution)?

Lande and Arnold (1983) also provide the equations for predicting changes in the variances and covariances between multiple traits due to natural selection. Given a set of m traits $(T_1, T_2, \ldots, T_m)$ measured on a collection of individuals, and a measure of relative fitness for each individual, one first regresses relative fitness (w) on the set of traits plus each possible binary product (e.g. $T_1^2, T_1T_2, \ldots, T_1T_m, T_2^2, T_2T_3, \ldots, T_m^2$). Next, one collects the regression coefficients into a $m \times m$ matrix (γ), called the "stabilizing selection gradient", in which the diagonal elements are the regression coefficients of the unitary terms ($T_1^2, T_2^2, \ldots T_m^2$) and the off-diagonal elements are the regression coefficients of the binary terms (T_iT_j). The change in the phenotypic variance–covariance matrix of the set of traits $\mathbf{T}=\{T_1, \ldots, T_m\}$ from before ($\mathbf{P}$) to after ($\mathbf{P*}$) the selection event is then: $\mathbf{P^*} = \mathbf{P}\gamma\gamma - (\mathbf{T^*} - \overline{\mathbf{T}})(\mathbf{T^*} - \overline{\mathbf{T}}) + \mathbf{P}.$

In essence, the multiple regression is estimating the adaptive surface relating fitness to traits. It is describing how natural selection determines the properties of the multivariate distribution of trait values. A multivariate normal distribution of traits (the most common assumption) requires only two types of parameters: a vector of means and a matrix of variances–covariances. Of course, if the relationship between relative fitness and traits follows some more complicated distribution then one would need to know the moments of the multivariate distribution. This provides a powerful graphic image: natural selection is the force that constrains the traits (and therefore the community-aggregated traits) to follow a particular distribution. The means, variances and covariances of functional traits in a given environment are the footprints of natural selection. Natural selection in a given environment imposes constraints on the means, variances and covariances of the traits and, at the community level, these are reflected in the community-aggregated traits.

Remember, from your introductory statistics course, that the variance (S_j^2) of trait j can also be written as: $S_j^2 = \overline{T_j^2} - (\overline{T_j})^2$. This means that you can define the community-aggregated variance of a trait by specifying two constraint equations. We do this by creating a "new" trait that is the square of the first one and is called the "second moment" of a distribution:

$$\overline{T_j} = \sum_{i=1}^{S} p_i T_{ij}$$

$$\overline{T_j^2} = \sum_{i=1}^{S} p_i T_{ij}^2.$$

In order to constrain both the community-aggregated mean at a value $(\overline{T_j})$ and the community-aggregated variance at a value (S_j^2), one would fix the second moment at $\overline{T_j^2} = S_j^2 + (\overline{T_j})^2$. It is also possible to use this trick to constraint higher moments.

Back to community dynamics

Okay. We have taken a rather lengthy detour away from community dynamics and into the arcane details of the Breeder's Equation. It's now time to link natural selection, via the Breeder's Equation, to community dynamics. This is done quite easily by combining Equation 5.3, relating the dynamics of relative abundances of genotypes in a community as a function of the traits of each species, with the Breeder's Equation that describes how the community-aggregated trait values change over time as a function of the selection pressure. In what follows I will use the univariate version of the Breeder's Equation for simplicity, but everything can be converted into the multivariate version as well. Remember that the univariate version of the Breeder's Equation is $\Delta \overline{T}(t+1) = h^2(t)S(t) = h^2(t)\left(\overline{T}^*(t) - \overline{T}(t)\right)$, where $\overline{T}(t)$ is the average trait value before selection and $\overline{T}^*(t)$ is the average trait value after selection. We can re-write this as $\overline{T}(t+1) = h^2(t)\sum_i T_i\left(p_i^*(t) - p_i(t)\right) + \overline{T}(t) = h^2(t)\sum_i T_i \Delta p_i(t) + \overline{T}(t)$, where $p_i(t)$ and $p_i^*(t)$ are relative abundances of species i before and after the selection event. The $\Delta p_i(t)$ term looks rather familiar since it is the change in the relative abundance of genotype i at time t. But we know from Equation 5.3 that the rate of change of the relative abundance of genotype i at time t is given[18] by $dp_i(t)/dt = p_i(t)\left[\beta(t)\left(T_i - \overline{T}(t)\right) + \varepsilon_i(t)\right]$. We can approximate the instantaneous change in p_i at time t (i.e. $dp_i(t)$), by the change in p_i during the small interval of time Δt (i.e. $\Delta p_i(t)\Delta t$), so long as the time interval (Δt) is very short. If we make this approximation and insert it into Equation 5.2 (Box 5.5) we get $\overline{T}(t+1) = \sigma_{T_0,T_p}(t)\beta(t) + h^2(t)\sum_i T_i\varepsilon_i(t) + \overline{T}(t)$. But this is simply the Breeder's Equation expressed in terms of fitness plus a random component $(h^2(t)\sum_i T_i\varepsilon_i(t))$! If we iterate this equation over time then we get:

---

[18] If the relationship between traits and per capita growth rates is nonlinear or interactive then the equation must be modified accordingly.

$$\overline{T}(t) = \sum_{j=0}^{t} h^2(j)\beta(j)\sigma_T^2(j) + \sum_{j=0}^{t} h^2(j)\sum_i T_i\varepsilon_i(j) + \overline{T}(0)$$

$$\overline{T}(t) = \sum_{j=0}^{t} \sigma_{T_0,T_p}(j)\beta(j) + \sum_{j=0}^{t} h^2(j)\sum_i T_i\varepsilon_i(j) + \overline{T}(0). \qquad \text{(Eqn. 5.6a,b)}$$

This equation is quite important for our purpose in explaining the origin of community-aggregated traits and in explaining why community-aggregated traits will constrain community assembly. It says that the value of the community-aggregated trait at any time t is determined by the relative strengths of three things: (i) the community-aggregated value at the start of community assembly $(\overline{T}(0))$ and representing what we would find in the absence of selection (i.e. a neutral expectation), (ii) the effects of random sampling effects due to finite population sizes $(\text{drift} : \sum_{j=0}^{t} h^2(j)\sum_i T_i\varepsilon_i(j))$ and (iii) the effects of selection on the trait $(\sum_{j=0}^{t} h^2(j)\beta(j)\sigma_T^2(j))$. When will the effects of selection on the determination of the community-aggregated trait disappear? When either there comes a time (j) when there is no variation in the trait $(\sigma_T^2(j))$, when there comes a time when there is no effect of the trait on per capita growth rate $(\beta(j))$ or when there comes a time when there is no heritability of the trait across generations $(h^2(j))$. Let's now turn to the part of the equation dealing with random population drift (i.e. the neutral part of the dynamics) and see how it will affect the community-aggregated trait at time j. We see, first, that if the trait is not heritable then drift will not change the expected community-aggregated trait. However, drift will not change the expected community-aggregated trait even if the trait is heritable. This is because the random changes in the per capita growth rates $(\varepsilon_i(j))$ due to sampling fluctuations are, by definition, independent of the properties (including the traits) of the species and have a mean of zero, therefore $\sum_i T_i\varepsilon_i(j)$ will be close to zero unless the population sizes and the species pool are both very small; if this happens then chance initial associations between a trait value and those few lucky species who increase due to drift can cause a systematic change in the community-aggregated trait. The community-aggregated trait value will fluctuate randomly around the initial value $\overline{T}(0)$ in the absence of selection. As long as there is a link between the heritable trait and the probabilities of survival and reproduction (thus fitness), the community-aggregated trait value will reflect natural selection over time as described by the Breeder's Equation plus such random sampling effects. The community-aggregated trait constrains the relative abundances because natural selection constrains which genotypes will increase or decrease over time.

Box 5.5. **Linking population dynamics and natural selection**

The equation for community dynamics (Equation 5.2) is

$$\frac{dp_i(t)}{dt} = p_i(t)[r_i(t) - \bar{r}(t)].$$

Relating the per capita growth rates at time $t(r_i(t))$ to traits that bias probabilities of survival, reproduction and immigration, and assuming for simplicity a linear relationship, the equation for community dynamics (Equation 5.3) is

$$\frac{dp_i(t)}{dt} = p_i(t)\left[\beta(t)(T_i - \bar{T}(t)) + \varepsilon_i(t)\right].$$

The univariate Breeder's Equation (Equation 5.5) is:
$$\bar{T}(t+1) = h^2(t)S(t) + \bar{T}(t) = h^2(t)\left(\bar{T}^*(t) - \bar{T}(t)\right) + \bar{T}(t).$$
We can re-write this as:
$$\bar{T}(t+1) = h^2(t)\sum_i T_i\left(p_i^*(t) - p_i(t)\right) + \bar{T}(t) = h^2(t)\sum_i T_i[\Delta p_i(t)] + \bar{T}(t).$$
Approximating the infinitesimal change $dp_i(t)$ of Equation 5.3 by $\Delta p_i(t)$ for one time step ($\Delta t = 1$), we substitute $\Delta p_i(t)$ of the Breeder's Equation by $dp_i(t)\,dt$ of the equation for community dynamics, giving:

$$\bar{T}(t+1) = h^2(t)\sum_i T_i\left[p_i(t)(\beta(t)(T_i - \bar{T}(t)) + p_i(t)\varepsilon_i(t)\right] + \bar{T}(t),$$

$$\bar{T}(t+1) = h^2(t)\left[\sum_i \beta(t)p_i(t)T_i^2 - \sum_i \beta(t)p_i(t)\bar{T}(t)T_i + \sum_i T_i\varepsilon_i\right] + \bar{T}(t).$$

Now, $\sum_i \beta(t)p_i(t)T_i^2 = \beta(t)\sum_i p_i(t)T_i^2 = \beta(t)\overline{T^2}(t)$ and
$\sum_i \beta(t)p_i(t)\bar{T}(t)T_i = \beta(t)\bar{T}(t)\sum_i p_i(t)T_i = \beta(t)\left(\bar{T}(t)\right)^2.$
Therefore, $\bar{T}(t+1) = h^2(t)\left[\beta(t)\overline{T^2}(t) - \beta(t)\left(\bar{T}(t)\right)^2 + \sum_i T_i\varepsilon_i\right] + \bar{T}(t).$

Finally, we note that the variance in the trait values at time t ($\sigma_T^2(t)$) is defined as $\sigma_T^2(t) = \overline{T^2}(t) - \left(\bar{T}(t)\right)^2$ and therefore
$\beta(t)\overline{T^2}(t) - \beta(t)\left(\bar{T}(t)\right)^2 = \beta(t)\left[\overline{T^2}(t) - \left(\bar{T}(t)\right)\right] = \beta(t)\sigma_T^2(t).$ This gives
$$\bar{T}(t+1) = h^2(t)\beta(t)\sigma_T^2(t) + h^2(t)\sum_i T_i\varepsilon_i(t) + \bar{T}(t).$$
However, $\beta(t)$ in our dynamic equation is the slope of the regression of the trait on the per capita growth rate ($r_i(t)$), which is also an estimate of fitness. So $h^2(t)\beta(t)\sigma_T^2 = \left(\sigma_{T_0,T_P}(t)/\sigma_{T_P}^2(t)\right)\beta(t)\sigma_T^2(t) = \left(\sigma_{T_0,T_P}(t)\right)\beta(t)$ which is the genetic covariance of the trait $\left(\sigma_{T_0,T_P}(t)\right)$ times the slope of

> ### Box 5.5. (Continued)
> the regression of the trait on the per capita growth rate ($\beta(t)$). Thus, we end up again with the Breeder's Equation, expressed in terms of fitness (i.e. Box 5.4), plus a random component relating to sampling fluctuations: $\overline{T}(t+1) = \sigma_{T_O, T_P}(t)\beta(t) + h^2(t)\sum_i T_i \varepsilon_i(t) + \overline{T}(t)$.

Genotypes, phenotypes, environments and heritability

Every scientific model chooses to emphasize certain aspects of reality while ignoring others; if it didn't then the model would be as complicated as reality itself and nothing would have been gained. In this sense, every scientific model is a lie and the model developed in this book is no exception. However, like successful politicians, a good model must be able to lie without getting caught. Those parts of reality that we leave out of our model cannot distort the model behavior so much that it becomes only a caricature of reality. In this book I refer to things called "species traits" and pretend that they are constant properties of species, but this is clearly a lie. I know that most morphological, physiological and phenological traits that are important to fitness both differ between genotypes of the same species and display plastic change with environmental conditions. Furthermore, since the species' traits are constant in the model, it assumes that evolution of traits within species is so small relative to the pre-existing trait variation between species during the time scale of community assembly that it can be ignored. Why should the model usually be able to get away with such lies, and when might it get caught?

Replacing genotype means with species means

Let's put off the questions related to trait plasticity and evolution for a moment and first look at the effect of replacing genotype trait means by species means when computing community-aggregated trait values. In our model I am sub-dividing this population of genotypes into different species. I will assign a unique index to each unique genotype (k) and to each species (i) in the community and write "$k \in i$" to mean "the set of genotypes belonging to species i", since the same genotype cannot exist in more than one species. The abundance (in terms of individuals or biomass) of genotype k of species i is a_{ik}, the abundance of all genotypes of species i is $a_i = \sum_{k \in i} a_{ik}$ and the total abundance of all genotypes in the community is $A = \sum_{i=1}^{S} a_i = \sum_{i=1}^{S} \sum_{k \in i} a_{ik}$. If the trait value of genotype k, which occurs in species i, is T_{ik} then the average trait value of all genotypes in the community is, by definition,

$$\overline{T} = \frac{1}{A}\sum_{k=1}^{K} a_{ik}T_{ik} = \frac{1}{A}\sum_{i=1}^{S}\sum_{k\in i} a_{ik}T_{ik} = \frac{1}{A}\sum_{i=1}^{S}\left(a_i\left(\frac{\sum_{k\in i} a_{ik}T_{ik}}{a_i}\right)\right).$$

The average trait value of species i is, by definition, $\overline{T} = \frac{\sum_{k\in i} a_{ik}T_{ik}}{a_i}$. It follows that $\overline{T} = \frac{1}{A}\sum_{i=1}^{S} a_i\overline{T}_i = \sum_{i=1}^{S} p_i\overline{T}_i$. So, what is the effect of replacing genotype means by species means in the Breeder's Equation? None, if the replacement is done correctly since the community-aggregated trait $\overline{T}$ can be obtained from the species means $\overline{T}_i$ without explicitly considering the intraspecific genotypic variation of trait values. However, for this to be true, the species' means must be obtained from an unbiased sample of the genotypes within each species and the community value must be obtained as a weighted average of the species' means, where the weighting is based on the relative abundance of each species in the community. If we randomly sample individuals from each species then this will give a representative sample of genotype frequencies. In particular, it is *not* correct to calculate the community-aggregated value as $\overline{T} = \frac{1}{S}\sum_{i=1}^{S} \overline{T}_i$ which implicitly (and almost always incorrectly) assigns equal relative abundances to each species.

Trait plasticity

The ecological notion of "plasticity" is not completely fixed and can mean slightly different things to different people. By "plasticity" I refer to the phenomenon in which the same genotype expresses different phenotypes in different abiotic and biotic environments. Some traits are so conservative[19] that they are constant across environments (and, because of this, are used by systematists to define phylogenetic levels) but most traits related to fitness are plastic to varying degrees. It would seem that any model pretending that traits are species' constants is telling a rather big lie.

Not if the lie is done properly. Remember that the MaxEnt model predicts relative abundances based on the differences between species in terms of their trait values. Two species having exactly the same traits would always have the same predicted relative abundances and the more different the traits

---

[19] Evolutionary biologists would call these "canalized" traits.

between two species, the more different the predicted relative abundances.[20] Therefore the prediction error in the model will depend on the plasticity of the trait values within a given species in different environments relative to the variation in the average trait values across species in the species pool. So what are the relative sizes of intraspecific and interspecific variation in functional traits? We can evaluate this question in three ways. First, we can look at the empirical evidence measuring the correlation of trait values across species, or the species ranks in the trait values, when these trait values are measured in different environments or in different years. Second, we can look at the empirical evidence measuring the percentage of variation in trait variation that can be attributed to interspecific and intraspecific differences and how these percentages relate to environmental differences (variance components). Finally, we will look at the theoretical relationship between heritability and each of three sources of variation when measured at the intraspecific and interspecific levels: genotypes, environments and genotype–environment interactions.

Components of variance in trait values

Let's be a bit more precise about what I mean by "sources of variation" in trait values and how these can be compared. Given any group of individuals in a community we can divide these individuals into different species, into different genotypes within each species and into different clones within each genotype (for plants; this level rarely exists in nature for animals). If T_{ijk} is the trait value of T for clone k of genotype j of species i, then we can decompose the difference between T_{ijk} and the mean value into the sum of three types of deviation: the deviation of the clones from the genotype mean $\left(T_{ijk} - \overline{T_{ij}}\right)$, the deviation of the genotype means from the species mean $\left(\overline{T_{ij}} - \overline{T_i}\right)$, and the deviation of the species means from the community mean $\left(\overline{T_i} - \overline{T}\right)$, as follows: $\left(T_{ijk} - \overline{T}\right) = \left(T_{ijk} - \overline{T_{ij}}\right) + \left(\overline{T_{ij}} - \overline{T_i}\right) + \left(\overline{T_i} - \overline{T}\right)$. Each of these types of deviation contributes to the overall variance and the contribution of each type to the overall variance of the trait is called a "variance component". To calculate these variance components we must specify an error structure for our trait; here I will assume that it is normally distributed and write "$N(0, \sigma^2)$" to mean "a normally distributed random variable having a mean of zero and a

---

[20] This is not strictly true since the traits are weighted by the Lagrange Multipliers but it is a useful rule of thumb.

variance of σ^2". We can write the following mixed model[21] (Pinheiro and Bates 2000, Hox 2002):

$$T_{ijk} = \mu_{ij} + \varepsilon_{ijk} \qquad \varepsilon_{ijk} = \left(T_{ijk} - \mu_{ij}\right) = N(0, \sigma^2_{ijk})$$

$$\mu_{ij} = \mu_i + \delta_i \qquad \delta_i = (\overline{T_{ij}} - \mu_i) = N(0, \sigma^2_{ij})$$

$$\mu_i = \mu + \varphi_i \qquad \varphi_i = (\overline{T_i} - \mu) = N(0, \sigma^2_i).$$

The three variance terms $(\sigma^2_{ijk}, \sigma^2_{ij}, \sigma^2_i)$ are called the "components" of the overall variance since their sum gives the total variance of the trait: $\sigma^2_{ijk} + \sigma^2_{ij} + \sigma^2_i = \sigma^2$. The σ^2_{ijk} term is the variation due only to differences in the trait within clones of the same genotype and represents pure environmental or ontogenetic effects. The σ^2_{ij} term is the added variation in the trait (beyond that of σ^2_{ijk}) due only to differences between genotypes of the same species and represents the intraspecific variation between genotypes. Finally, the σ^2_i term is the added variation of the trait (beyond that of σ^2_{ijk} and σ^2_{ij}) due only to differences between genotypes of different species, and so represents the interspecific variation. The model can be expanded to include environmental effects and genotype–environment interactions as well as other levels of relatedness of genotypes within species and is an important tool in quantitative genetics and plant and animal breeding. Since plant and animal breeders are interested in *breeding* (thus trait evolution) the model is almost never extended to the interspecific level. This, however, is owing to a lack of interest, rather than of possibility.

An example from my own research can make this notion more concrete, although the experiment was not specifically designed for this purpose and so did not include the within-individual (clonal) level. Specific leaf area (*SLA*, projected leaf area divided by leaf dry mass, $cm^2 g^{-1}$) is an important functional trait of plants because it affects many aspects of plant performance. It is a component of plant relative growth rate (Blackman and Wilson 1951, Poorter and Remkes 1990), it affects leaf palatability to herbivores (Elger and Willby 2003), it affects maximum net photosynthetic rate (Lambers and Poorter 1992, Reich *et al.* 1997, 1998) and also leaf nutrient economy through its effect on leaf lifespan (Aerts 1990, Reich *et al.* 1992, Diemer 1998). *SLA* is also a plastic trait that generally increases in leaves that develop in low light levels or when the plant is growing in limiting nutrient levels. For example, I (Shipley 2000b) have documented a change in *SLA* of 44%

---

[21] The parameters of this model are estimated using maximum likelihood techniques. Mixed models can be fit using most modern statistical packages.

(*Lythrum salicaria*) and 47% (*Epilobium glandulosum*) within six hours following a change in light levels in the same genotypes of two particularly plastic species. However, it is also obvious, sometimes painfully so,[22] that different species have different typical *SLA* values and that *SLA* varies across environmental gradients.

Driss Meziane and I (1999) were interested in understanding how *SLA* changes in a non-genetic (plastic) manner across environments and how it changes genetically within and between species. Driss Meziane (1998) randomly allocated individuals of each of 22 species of herbaceous plants in each of four different controlled environments (high/low light intensity (100 vs. 200 μmol m^{-2} s^{-1} PAR) crossed with fertile/infertile nutrient supply (1 vs. 1/6 dilution of a standard hydroponic solution) and measured the average *SLA* of each plant (g cm^{-2}). The species' mean *SLA* values ranged from 77 to 475 cm^2 g^{-1}. Each plant was grown from seed and was presumably a different genotype. The species were all herbaceous species typical of unshaded upland habitats of north-eastern North America and so they certainly did not span the entire range of interspecific variation in *SLA*. In fact, judging from the worldwide data set of Wright *et al.* (2004) this set of species represented only about 50% of the worldwide interspecific range. The total variance in *SLA* was 8865.8 and this variance was decomposed into those components that were the result of (i) interspecific differences, (ii) to differences between environments, (iii) to differences between species in the way that *SLA* changed in response to environments (a "species–environment" interaction) and (iv) to differences in *SLA* between genotypes (individuals) within a given species and in a given environment. Table 5.2 summarizes the results.

The last term ("individuals") gives the variation between individual plants of the same species growing in the same average environment. This value (16.6% of the total variation) is a pure genetic component and quantifies the genetic variation at the intraspecific level under constant environmental conditions. The first term ("between species") gives the *added* genetic variation beyond that due to the fact that we are comparing different individuals (39.9% of the total variation) that results when comparing individuals growing in the same environment if these individuals belong to different species. Together these two components (16.6 + 39.9 = 56.5%) quantify the degree to which differences in phenotypic values of

---

[22] While botanizing in the hills inland from Alicante (Spain), a region with typical Mediterranean vegetation, I could not walk without being cut, stabbed and scratched because most plant species had such sclerophyllous leaves and such low *SLA*.

Table 5.2. *Variance components (column 2) and total variance (column 3) of specific leaf area.*

Source of variation	Variance unique to this level (i.e. variance components)	Percent of the total variance
Between species	3537.3	39.9
Between environments	2592.7	29.2
Species × environments	1265.6	14.3
Individuals	1470.2	16.6

SLA are attributable to purely genetic differences between genotypes. The second term ("between environments") gives the variation between individual plants of the same species growing in the different environments. Because the individual plants were not clones, this is not exactly a purely environmental component of the phenotypic variation. However, because the individuals were randomly assigned to the different environments, there should not be any *systematic* differences between genotypes. This value (29.2% of the total variation) is a non-genetic component and is entirely attributable to phenotypic plasticity. The term called "species × environments" is a hybrid entity that combines both genetic and environmental effects. It quantifies how much of the total variation is due to differing degrees of phenotypic plasticity among the different species. In other words, by how much the same environmental change will induce different changes in *SLA* due to the fact that the individuals come from different species; this is still a genetic component. We can see that differences between genotypes are responsible for 70.8% (i.e. 16.6 + 39.9 + 14.3) of the differences in the phenotypic values of *SLA*.

Remember why we are asking questions about intraspecific vs. interspecific vs. environmental trait variation. We want to know when we can get away with the lie that trait variation within species (thus intraspecific variation and trait plasticity within species) is sufficiently small relative to interspecific variation that we can replace all intraspecific trait values by a species mean. In part this is an empirical question that requires studies such as those reviewed above. However, we can go a long way towards an answer by looking at how our measure of heritability (h^2) is affected by these factors. This is important because the heritability determines how much of the difference between genotype traits and community-aggregated traits is translated into changes in relative abundance. The next section will explore this question from a more theoretical viewpoint.

Genetic and environmental components
of heritability (h^2)

Differential survival and reproduction among parents that is associated with our trait will result in changes in its mean value; this is the selection gradient. The degree to which the change in the trait mean is transmitted to the next generation is determined by the heritability (h^2) and this will affect both the evolution of the trait among genotypes of the same species and the relative abundance of species. However, "heritability", the slope of the parent–offspring regression for the trait, is a statistical property that depends on the group of genotypes over which the regression is applied, not some fixed genetic property of a single genotype.[23] What factors can affect the value of h^2? How might such effects be expressed at an intraspecific level, thus affecting evolution, and at the interspecific level, thus affecting community structure?

In order to answer this question we must first take a closer look at the sources of variation in traits. All of the trait variation that we have considered so far concerns phenotypic variation. Let's call the phenotypic (i.e. trait) values T. Quantitative geneticists (Falconer and Mackay 1996) separate this phenotypic variation into four components: (i) differences between individuals due to differences in genotypes (G), (ii) differences due to non-genetic ("environmental") causes (E), (iii) differences due to genetic plasticity (i.e. when the same degree of change in the environment induces different degrees of change in phenotypes between genotypes), and (iv) the statistical interaction (I_{GE}) between the genotypes and environments representing (if present) a tendency for particular genotypes to be found in particular environments (a sort of non-genetic heritability). The total phenotypic variance of a trait (σ_T^2) is therefore composed of these sources of variation: $\sigma_T^2 = \sigma_G^2 + \sigma_E^2 + 2\sigma_{G,E} + \sigma_{G,E}^2$. Those differences between phenotypes that are caused by purely non-genetic factors (σ_E^2) cannot be transmitted to offspring and so are not responsible for either natural selection or community dynamics. Those differences between phenotypes that respond to non-genetic factors, but for which the quantitative value of the response is affected by underlying genotypes (the genotype–environment interaction, or $\sigma_{G,E}$) can be genetically transmitted to offspring and so can respond to natural selection. Those differences arising from purely environmental effects but that are *consistently* associated with different

---

[23] In fact, the heritability can even vary depending on the age structure of the population.

genotypes $\left(\sigma^2_{G,E}\right)$ cannot affect evolution but can still affect community dynamics if these genotypes are grouped into different species. The variability in the underlying genotypic values $\left(\sigma^2_G\right)$ is measured by comparing across genotypes growing in a constant environment; this can be further divided into nested series of relatedness (families, local populations, species and so on). The variability in the non-genetic effects $\left(\sigma^2_E\right)$ is measured by comparing the same genotype growing in the different environments and represents the average degree of environmental plasticity in the trait. Finally, the genotype–environment covariance is what is left over and measures the degree to which different genotypes differ in their environmental plasticity. Because it is usually not possible in animals to clone the same genotype to produce genetically identical individuals, researchers will randomly allocate different genotypes to different environments and then statistically "control" for genotypes.

We can show this using a path diagram (Figure 5.6) and use the rules of structural equations modeling to express the heritability (h^2) in nature as a function of these different sources of variation; see Shipley (2000a) for an extended discussion of this statistical method. The parental trait phenotype (T_p) is determined by the midparental genotype G_p plus non-genetic causes: $T_p = G_p + E_p$. Correlations between the genetic and non-genetic causes of

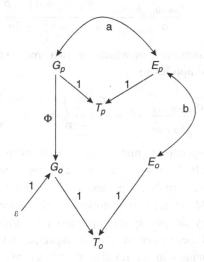

Figure 5.6. Path diagram showing the causal (single-headed arrows) and non-causal (double-headed arrows) relationships between genotypes (G), environments (E) and traits (T) of parents (P) and offspring (O).

the phenotype are quantified by a free covariance (a) and this represents both genotype×environment interactions and the fact that particular genotypes might be preferentially found in particular environments (niche partitioning). The value of the path coefficient linking the parental and offspring genotypes (Φ) is the relationship that *would* be found between the trait values of offspring and parents if the relationship was measured in a perfectly constant environment, thus removing all non-genetic causes of trait variation; the variation in offspring genotypes from the same parents is given as ε. The value of ε is the "segregation" variance of quantitative genetics if all offspring of a parental pair experience exactly the same environment. The value of Φ would be unity if there were no genetic mutations in the offspring and so will be only slightly lower than unity in nature (owing to rare mutational changes). This value will not change much across genotypes. The genetic variance of such offspring that are conceived by the same parents and are born and raised in the same environment $(\sigma^2_{G_o})$ can be partitioned as $\sigma^2_{G_o} = \sigma^2_{G_p} + \sigma^2_\varepsilon$. Finally, we allow the non-genetic causes that exist in the parental (E_p) and offspring (E_o) generations to be potentially correlated as well (b); this quantifies the degree to which parents and offspring tend to occur in the same environments.

Now, the heritability, as measured in the field and derived from the rules of path analysis, is defined as

$$h^2 = \frac{\sigma_{parents,\ offspring}}{\sigma^2_{parents}} = \frac{\Phi\left(\sigma^2_{G_p} + a\right) + b}{\sigma^2_{G_p} + \sigma^2_{E_p} + 2a}.$$

In a controlled experiment, in which the two free covariances (a and b) would be zero, this simplifies to

$$h^2 = \frac{\sigma_{parents,\ offspring}}{\sigma^2_{parents}} = \frac{\Phi\left(\sigma^2_{G_p}\right)}{\sigma^2_{G_p} + \sigma^2_{E_p}} \approx \frac{\left(\sigma^2_{G_p}\right)}{\sigma^2_{G_p} + \sigma^2_{E_p}}.$$

This equation is important to understanding heritability at the intraspecific and interspecific levels. It says that as the genotypes become more different (as $\sigma^2_{G_p}$ increases), the heritability increases. If the genotypes were identical the heritability would be zero. If genotypes within species resemble each other more than they do genotypes separated into different species, then the heritability will be larger at the interspecific level (where natural selection leads to changes in the relative abundance of species) than at the intraspecific level (where natural selection leads to changes in the relative abundance of different genotypes of the same species – i.e. evolution). Since the non-genetic (environmental) variation $\left(\sigma_{E_p}^2\right)$ of the trait – that is, pure trait plasticity – will not change systematically when going from the intraspecific to the interspecific level then, the more different the species, the closer the heritability will be to unity. Heritability determines the degree to which the selection gradient experienced by parents is translated into a change in average trait value of the offspring. Since this determines both evolution at the intraspecific level and changes in community structure at the interspecific level, this means that community structure will change more quickly than will the evolution of trait differences within species.

So, since interspecific variation in traits is usually much larger than intraspecific trait variation, and since the heritability of traits is usually stronger at the interspecific level than at the intraspecific level, we should be able to ignore intraspecific genotypic variation in our model of community assembly by treating all genotypes of a species as having the same (average) trait value without introducing too much error. Of course, this conclusion is quite meaningless with respect to evolution since, for evolution, interspecific variation in genotypes is meaningless.

Thus the main message of this chapter: natural selection of genotypes within the same species leads to evolution. Natural selection of genotypes between different species leads to population dynamics and community assembly. Between species, the consequence of natural selection is to force community-aggregated traits to follow specific trajectories in time; that is, to *constrain* community-aggregated traits to values that they would not otherwise take in the absence of natural selection. This is why knowledge of community-aggregated traits provides *information* about the relative abundance of species having different trait values. It is why, if we knew the values of all of the community-aggregated values of the traits responding to natural selection, only random sampling variation would be left. This, in turn, is why maximizing the relative entropy subject to all of the relevant community-aggregated traits is equivalent to choosing the most likely distribution of what are random allocations, once the systematic biases have been taken into account.

Natural selection, community-aggregated traits and the predictive ability of the Maximum Entropy model

If you are a community ecologist then you must feel a bit punch-drunk after so much population genetics so let's pause and take stock. In this chapter we have found a theoretical link between population dynamics, natural selection and community-aggregated traits. In the previous chapter we found a theoretical link between community-aggregated traits, the traits of each species in the species pool, and the predicted relative abundances of each species thanks to the Maximum Entropy Formalism. In the next chapter we will look at some empirical data to see how well we can predict community structure via community-aggregated traits but that example will not explicitly include dynamics. Ideally we should look at some empirical data in which each component (traits, dynamics, natural selection, community-aggregated traits and predicted relative abundances) is measured. We would seed out an initial "community" from a species pool in which we would have measured all of the important functional traits that are responding to natural selection at the site. We would then compare observed and predicted relative abundances during community assembly in this experiment. By manipulating various parameters (the distribution of traits, the number of traits, the number of species, the effects of small initial population sizes, the strength of selection and so on) we could then get an idea of how such things affect the predictive ability of the model.

Unfortunately, such an experiment still waits to be done. On the other hand, it costs relatively little to simulate such "experiments" on a computer and so I will do this as a partial solution until the proper experiments have been conducted. Although there are several advantages of using such simulations (besides the one to my research budget) one obvious disadvantage is that any detailed results will depend on the accuracy of the model of population dynamics that is used. Since I spent much of Chapter 2 arguing that we don't really have such a realistic model, it makes little sense to put too much faith in the detailed results of any particular simulation. However, by incorporating at least some of the most basic dynamic properties into a simple dynamic model, we can still get a feeling for the qualitative effects of different parameters of the dynamics and how these are likely to affect the predictive ability of the model. I'll first describe the dynamic model that will be used in the simulations and then return to Equation 5.3 to help interpret the results.

We start with a mathematical necessity. It is always true that the number of individuals of species i one time interval into the future $(t + 1)$ equals the number of individuals of this species at time t, plus the additions of new individuals during the time interval, minus the losses of existing individuals during the time interval. If $b_i(t)$ is the average number of new individuals added per existing individual of species i at time t, and $d_i(t)$ is the probability of an existing individual of species i at time t dying or emigrating, then our mathematical necessity is $n_i(t + 1) = n_i(t)(1 + b_i(t) - d_i(t))$. Let's first assume a closed population (no immigration or emigration) so that $b_i(t)$ and $d_i(t)$ represent birth and death rates. The per capita rate of growth is the difference between births and deaths: $r_i(t) = b_i(t) - d_i(t)$ and so our mathematical necessity becomes $n_i(t + 1) = n_i(t)(1 + r_i(t))$. Now, since $r_i(t)$ is determined by the process of birth and death of the individuals in the population, this value would be very stable in a very large population and be very close to the theoretical probabilities of birth and death at time t. However, in any finite population there will be random variation of actual rates of birth and death. We can therefore decompose the actual observed value of $r_i(t)$ into a theoretical value $\hat{r}_i(t)$, that holds in an infinitely large population, plus a random component $\varepsilon_i(t)$ representing the fluctuations around this theoretical value due to sampling from a finite population; this gives $n_i(t + 1) = n_i(t)(1 + \hat{r}_i(t) + \varepsilon_i(t))$. The population dynamic will be complete once we decide how to model $\hat{r}_i(t)$ and $\varepsilon_i(t)$. In these simulations I will model the random component as a normally distributed variable with a zero mean and a standard deviation of $\sigma/\sqrt{n_i(t)}$.

The dynamic model that I will use (Equation 5.7) is simply the discrete version of the Lotka–Volterra equations (Chapter 2) with two additions. In the Lotka–Volterra equations,

$$\hat{r}_i(t) = r_i^* \left(\frac{K - \sum_{j=1}^{S} a_{ij} n_i(t)}{K} \right).$$

I first want to add an immigration term representing new individuals of species i arriving into the population from the species pool: $m_i(\lambda_i, t)$. In all of the simulations that follow $m_i(\lambda_i, t)$ is a Poisson random variate with a species-specific mean per capita immigration rate of λ_i. Note that this rate does not depend on the actual population size of species i in our local community since individuals are coming from outside, although $m_i(\lambda_i, t)$ will normally depend on the population size of species i in the larger metapopulation even if the probability of immigration per individual is the same (the assumption of neutral models). Second, I want to add a loss rate representing random density-independent disturbances that kill off a certain proportion of all individuals in the community. If the probability of an individual being killed by disturbance[24] at time t is $d(t)$ then the number of individuals of species i that are lost in this way at time t is $d(t)n_i(t)$. I want this disturbance rate to represent mortality events that only occur occasionally but that, when they

---

[24] I prefer not to subsume this source of mortality into the per capita growth rate because, as you will see, this disturbance rate will be something that occurs only rarely and with different intensities.

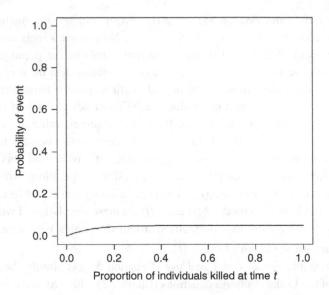

Figure 5.7. There is a 95% chance that no density-independent mortality event occurs. If it does occur, the proportion of all individuals who die is given by a beta distribution with shape parameters of 1 and 10.

occur, can be substantial. I therefore model $d(t)$ as a Bernoulli random variable with a probability of 0.05 (the probability of a disturbance occurring) times a beta random variable with shape parameters of 1 and 10. Thus, on average, a disturbance event occurs once in every 20 time intervals and, when it occurs, it is usually small in magnitude but occasionally can be large. Figure 5.7 shows the distribution of such events.

This model (Equation 5.7) has the virtue of being well-known, simple and still containing the basic properties of any population dynamics: the potential for exponential growth, a negative feedback with increasing numbers of individuals, a random drift component, unpredictable and rare disturbances, and immigration. The system of discrete difference equations for the S species in the species pool is:

$$n_i(t+1) = n_i(t)(1 + r_i(t) + \varepsilon_i(t)) - d(t)n_i(t) + m_i(\lambda_i, t)$$

$$n_i(t+1) = n_i(t)\left(1 + r_i^*\left(\frac{K - \sum_{j=1}^{S} a_{ij}n_j(t)}{K}\right) + \varepsilon_i(t)\right) - d(t)n_i(t) + m_i(\lambda_i, t).$$

(Eqn. 5.7)

In this equation $n_i(t)$ is the population size of species i at time t. There are three species-level properties: (i) r_i^* is the maximum per capita growth rate of species i, (ii) a_{ij} is the amount by which one average individual of species j reduces the per capita growth rate of species i (i.e. the "competition coefficient") and (iii) λ_i is the average number of new individuals of species i that immigrate into the site at each time interval.

In order to include traits into Equation 5.7, I allow these three species-specific parameters to be functions of these traits. A common scenario of community assembly during secondary succession is one in which there is a tradeoff between the ability to reproduce and disperse rapidly in the absence of competition and competitive ability. Thus, those species that have the highest maximum per capita growth rates and immigration abilities have the lowest competitive abilities. To represent this tradeoff in the non-neutral case, I make the values of r_i, a_{ij} and λ_i be functions of the K traits. Specifically:

$$r_i = \frac{0.15}{\bar{r} \sum\limits_{k=1}^{K} w_{1k}} \sum_{k=1}^{K} w_{1k} k T_{ik}^{w_{1k}}$$

$$a_{ij} = \frac{1}{\sum\limits_{k=1}^{K} w_{2k}} \sum_{k=1}^{K} \left(w_{2k} \frac{T_{jk}}{T_{ik}} \right)$$

$$\lambda_i = \frac{1}{\sum\limits_{k=1}^{K} w_{3k}} \sum_{k=1}^{K} w_{3k} T_{ik}. \qquad \text{(Eqn. 5.8)}$$

The values of w_{1k}, w_{2k} and w_{3k} are weights that affect the amount by which each species-specific demographic parameter is affected by each of the traits. By varying the value of these weights we can simulate different strengths of selection on each trait. In the case of a purely neutral model I will use fixed values of $r_i = 0.1$, $a_{ij} = 1$ and $\lambda_i = 0.05$ for all species.

Before introducing the simulation results I want to go back to Equation 5.3. This equation $\left(dp_i(t)/dt \right) = p_i(t) \sum_k \beta_k(t) \left(T_{ik} - \bar{T}_k(t) \right) + \left(p_i(t) \varepsilon_i(t) \right)$ can suggest the sort of things that will affect predictive ability since it tells us by how much the actual relative abundance of species i at time t (i.e. $p_i(t)$) will change over the next instant in time. It includes four components: (i) the relative abundance of species i at time t ($p_i(t)$), (ii) the strength of natural selection acting on each trait at time t (the $\beta_k(t)$), (iii) the difference between the traits of species i and the community-aggregated traits $\left(T_{ik} - \bar{T}_k(t) \right)$ and (iv) the effect of random population drift ($\varepsilon_i(t)$).

Strength of selection versus drift

How will variation in the strength of natural selection affect the predictive ability of the Maximum Entropy Formalism in the context of our model? Recall that the $\beta_k(t)$ values in Equation 5.3 are the partial regression slopes of each of the k traits on the actual per capita growth rates of the species in the species pool at time t and $\varepsilon_i(t)$ is the deviation of the per capita rate of population change of genotype i from the expected value at time t due to random sampling variation (i.e. population drift). The first thing to notice is that if all $\beta_k(t)$ are zero (i.e. if there is no relationship between traits and per capita growth rates) then Equation 5.3 reduces to $dp_i(t)/dt = p_i(t)\varepsilon_i(t)$. This is a neutral model for which the dynamics of all species are completely determined by the drift component. If this is the case then the Maximum Entropy Formalism would have no predictive ability. If all of the $\beta_k(t)$ values are large then the effect of natural selection $\left(\sum_k \beta_k(t)\left(T_{ik} - \overline{T}_k(t)\right)\right)$ will dominate the effect of drift $(\varepsilon_i(t))$ and the predictive ability will be high.

To explore the effect of increasing the strength of natural selection, I first simulate Equations 5.7 by using a species pool of 25 species over 500 time intervals ("generations"). I assign a single trait value to each of the 25 species and this value is drawn from a normal distribution with mean of 100 and a standard deviation of 30. In the neutral case, in which traits are independent of demographic parameters (i.e. no natural selection), I assign $r_i^* = 0.15$, $a_{ij} = 1$ and $\lambda_i = 1$ for all species irrespective of their trait values. Each species begins with 20 individuals and the drift component has a standard deviation of 0.01. Figure 5.8 shows a typical simulation run. There is no relationship between the abundance of a species at time 500 and its trait value. The community-aggregated value of the trait fluctuates randomly over time with no systematic tendency, and stays close to the average value of 100. Most importantly, the bias-corrected r^2 value[25] between the actual relative abundances at time 500 and those predicted from the Maximum Entropy Formalism, given the community-aggregated trait value at time 500 and the traits of each species in the species pool, is essentially zero. To get an idea of the range of r^2 values that we might expect, I repeated the simulation 100 times; the median value[26] was -0.03 and 75% of the values were between -0.04 and 0.01.

Now, let's allow the demographic parameters to be functions of the trait, as specified in Equation 5.8. I will let w_1 be 0.6 so that $r_i^* \propto T_i^{0.6}$, meaning

---

[25] See Chapter 4; this corrects for bias due to the number of species in the species pool.

[26] No, this is not an error. The bias-corrected r^2 corrects for the expected r^2 value and so the actual value can be less than zero.

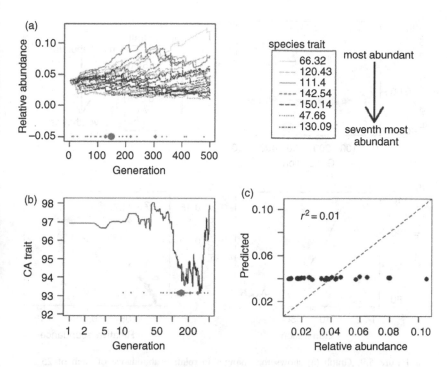

Figure 5.8. Graph (a) shows the changes in relative abundance of each of 25 species over 500 "generations" or time intervals when these species obey neutral dynamics. The top right panel lists the seven most abundant species and their trait values. Graph (b) shows the value of the community-aggregated value of this trait over time; note the logarithmic scale of time. Graph (c) shows the observed relative abundances at time 500 versus the predicted relative abundances given the community-aggregated trait value and the species' trait values, using the Maximum Entropy Formalism.

that species with larger values of the trait have higher maximum per capita growth rates. I will let w_2 be 0.5 so that $a_{ij} = 0.5(T_j/T_i)$, meaning that if species j has a larger trait value than species i then species j is the better competitor (i.e. species j has a stronger competitive effect on i than i has on j). Finally, I will let $w_3 = 0.01$ so that $\lambda_i = 0.01T_i$, meaning that species with larger trait values have, on average, more new immigrants arriving into the site at each time interval. Thus species with larger trait values have a higher potential per capita growth rate and a higher immigration rate but a lower competitive ability. The process of community assembly begins with only one individual per species and the drift variance is $0.01/\sqrt{n_i(t)}$. Figure 5.9 shows a typical simulation run. After 100 replicate runs, the median r^2 value was 0.84 and 75% of them were between 0.78 and 0.91.

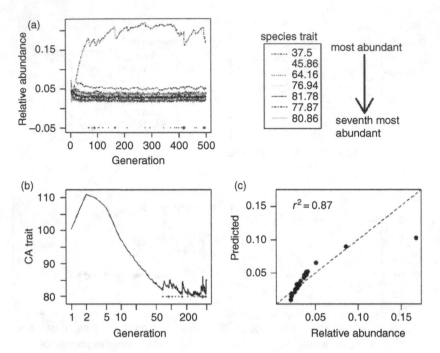

Figure 5.9. Graph (a) shows the changes in relative abundance of each of 25 species over 500 "generations" or time intervals when these species have demographic parameters that are functions of a single trait. The gray dots indicate disturbance events and their size is proportional to the proportion of individuals killed. The top right panel lists the seven most abundant species and their trait values. Graph (b) shows the value of the community-aggregated value of this trait over time; note the logarithmic scale of time. Graph (c) shows the observed relative abundances at time 500 versus the predicted relative abundances given the community-aggregated trait value and the species' trait values, using the Maximum Entropy Formalism.

Look at graphs (a) and (b) of Figure 5.9. For the first few time intervals there are very few individuals and so the populations of those species having the highest r_i^* values and immigration rates increase most rapidly. At this stage there is positive selection for rapid potential per capita growth rates and immigration abilities. Since these two demographic parameters increase with increasing values of the trait, this means that there is also a selection for increasing values of the trait. As a consequence, the community-aggregated trait value increases. However, as population sizes increase the actual per capita growth rates begin to slow down because of competition. Those species having weaker competitive abilities will have their actual per capita growth rates decrease most rapidly. Since higher trait values give weaker competitive abilities, this means that selection begins to favor species with smaller trait values. Thus, after about four time intervals the community-aggregated trait value begins to decrease and continues to decrease until about time interval 50. After this point the community-aggregated trait value stabilizes and then fluctuates randomly around a constant value. This is because, at this point, the community has reached a dynamic equilibrium with respect to the prevailing disturbance regime. The predictive ability of the Maximum Entropy Formalism is high but not perfect even though we have information on all relevant selection pressures (i.e. the single trait for each species and the resulting community-aggregated trait value). There is more to say about what happens when a dynamic equilibrium is reached, and the reasons for a less than perfect predictive ability even when we have perfect information on all relevant traits, but we will tackle these questions later.

At this point, let's explore the relationship between the strength (i.e. the bias-corrected r^2) of the relationship between traits and demographic parameters and predictive ability of the model. Obviously, the quantitative values of r^2 won't tell us much because they will depend on the specific details of the simulation model but we can at least get a qualitative idea. In order to simplify things, I will let only one demographic parameter be a function of the trait value. In the following simulations I will make $r_i^* = 0.15$ and $a_{ij} = 1$ for all species and independent of the trait values and the drift standard deviation will still be $0.01/\sqrt{n_i(t)}$ but the immigration rate will be a function of the trait values: $\lambda_i = w_3 T_i$. By varying the value of w_3 (the degree to which the trait affects the immigration rate) we can see how the predictive ability varies. In fact, even when $w_3 = 0.0001$, thus a very weak relationship between the trait values and the immigration rate, and when the drift standard deviation is $0.4/\sqrt{n_i(t)}$ the median r^2 was still 0.95. I obtained similar results when changing the weights for other demographic parameters. Thus, as soon as there is something for natural selection to work on (at least in this dynamic model) then the Maximum Entropy Formalism will use this information to give good predictions.

In Equation 5.3 there are two things that can affect the systematic component of the proportional rate of change of relative abundance due to natural selection $\left(\sum_k \beta_k(t)\left(T_{ik} - \overline{T_k}(t)\right)\right)$: the strength of the relationship between the traits and the demographic parameters and the variation of these trait values. Let's see what happens when we restrict the variation of the trait values. In the first non-neutral scenario the trait values were assigned to species from a normal distribution with a mean of 100 and a standard deviation of 30 and so 95% of species will have trait values between about 40 and 160. By changing

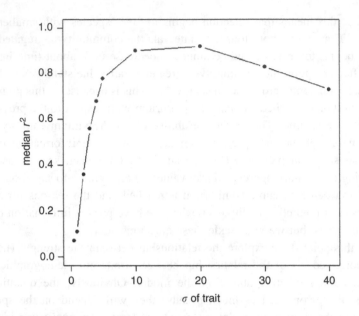

Figure 5.10. The median bias-corrected r^2 value of observed and predicted relative abundances for different levels of interspecific variation of trait values as a function of the interspecific variation in trait values. This simulation is based on a species pool of 25.

the standard deviation I can increase or decrease the variability of the trait values between species. When I repeat the same non-neutral scenario but change the standard deviation of the traits between species, each time running 100 simulations, I get the results shown in Figure 5.10.

Clearly, when species have very similar trait values, i.e. when $\sigma \approx 0$, then the predictive ability is reduced but this is to be expected given the term $\sum_k \beta_k(t)\big(T_{ik} - \overline{T}_k(t)\big)$. In the limit, when all trait values are exactly equal, then all traits are equal to the community-aggregated value, thus $\big(T_{ik} - \overline{T}_k(t)\big) = 0$ for all species and traits, and we recover a purely neutral model. The $\beta_k(t)$ values and the trait variation interact, as the equation shows, and so it is actually the combined effect of trait variation and the strength of the relationships between the traits and the demographic parameters (the $\beta_k(t)$ values) that is important. In the dynamic model that we are simulating it appears that only a modest amount of trait variation is required, even with weak $\beta_k(t)$ values, but this is specific to the particular dynamic model chosen. In fact, it is the ratio of the variation in the weighted trait values ($\beta_{k(t)}T_{ik}$) to the variation caused by drift ($\varepsilon_i(t)$) that is critical but, since the former will almost always be much bigger than the latter except at very small population sizes, any appreciable trait variation that affects demographic parameters should insure good predictive ability.

One curious part of Figure 5.10 is that the predictive ability actually begins to decrease above a certain level of trait variation. The explanation for this is rather different. When trait variation is too large then some species have traits that are so poorly adapted that they are quickly and completely excluded. This occurs both because selection is relentlessly pushing them out and also because they have so few individuals that random drift becomes important and hastens their demise even in the face of continuous immigration. Remember that the Maximum Entropy Formalism will always assign a non-zero probability (i.e. a predicted relative abundance) to every species in the species pool; assigning a Bayesian probability of zero means that it is logically impossible for a species to be found and therefore that it is not a member of the species pool. Now, if many of these poorly adapted species are actually extinct in the local community then this means that we will necessarily overestimate their predicted relative abundance (it must be greater than zero) and therefore necessarily underestimate the predicted relative abundance of the few dominant species that still survive (since the total relative abundance of all species must sum to unity). To take an extreme example, imagine a pool of 500 species but in which 498 of these species have traits that are so poorly adapted that they go extinct even though the Maximum Entropy Formalism predicts very small positive relative abundances of 0.001 for each. The remaining two species have identical trait values and the observed relative abundances of these two remaining species are (say) 0.4 and 0.6. The prediction error on each of the 498 "missing" species would be low (0 versus 0.001) but the *cumulative* predicted relative abundance of these 498 species would be large (0.498). Since the remaining two species have identical trait values their predicted relative abundances would be equal but, since the remaining total relative abundance is only 0.502 (i.e. 1–0.498), their predicted values would be only 0.251 each. As an example, look at Figure 5.11 which shows the result of one simulation using the same simulation setup as the first non-neutral scenario except that I include 500 species in the species pool.

The poorer predictive ability of the Maximum Entropy Formalism in Figure 5.11 is indeed an error of prediction but it doesn't arise from some error in the theoretical model but rather from the fact that we are sampling from a finite population of individuals and one cannot have less than one individual. For instance, in our extreme scenario above the predicted relative abundances of the 498 species were 0.001. If our vegetation sample only contains 100 individuals (say a $0.25\,\text{m}^2$ quadrat of herbaceous vegetation)

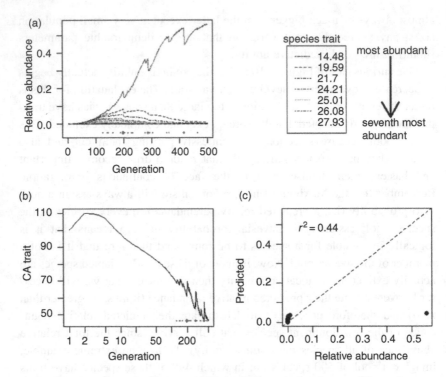

Figure 5.11. Same simulation as in Figure 5.10 except that there are 500 species in the species pool rather than 25.

then almost all such rare species would be missing $(0.001 \times 100 = 0.1$ individuals). If we could increase our sample size to include 10 000 individuals then even such rare species would still have some individuals and the prediction bias would be greatly reduced. Of course, as a practical matter, in order to include 10 000 individuals we might have to sample from such a large area that there is substantial environmental variation (and different selection pressures) within the sample and this would reduce our predictive ability as well.

We should expect such a decrease in predictive ability whenever there are many species included in the species pool that cannot actually persist in the vegetation. In particular, we should expect this decrease in predictive ability as the size of the species pool increases, as the strength of selection increases and as the interspecific variation in functional traits increases since we would be adding more and more species that cannot actually survive in the site. To explore this factor, I again ran the same simulations as in our first non-neutral case but varied the size of the species pool and the variance of the trait values.

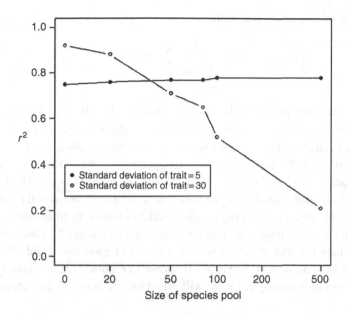

Figure 5.12. The effect of changing the size of the species pool and the inter-specific variation in the functional trait on the bias-corrected r^2 value between observed and predicted relative abundances provided by the Maximum Entropy Formalism.

Figure 5.12 shows the result. As expected, when there is wide trait variation between species, so that the strength of natural selection is strong and many species are pushed close to extinction, the result of increasing the number of species in the species pool is to reduce predictive ability for the reason given above. When there is only modest trait variation between species then this phenomenon is not seen even with a species pool of 500.

The above results are based on a scenario in which we have measured all traits that affect demographic parameters and that can therefore respond to natural selection. In reality we can never know this. What happens if we are ignorant of some of these traits? What is the consequence of ignoring traits whose relationship to the demographic parameters is strong or weak?

Intuitively, we would expect the predictive ability to decrease but let's explore this in a series of simulations. Since there are potentially three demographic parameters that are linked to traits (r_i^*, a_{ij}, λ_i) I will assign the same values of the first two parameters to all species independently of their trait values (thus, $r_i^* = 0.15, a_{ij} = 1$) and allow only the third parameter (the average number of immigrants per species per time interval) to vary as a function of a set of $K = 2$ traits:

$$\lambda_i = \frac{1}{\displaystyle\sum_{k=1}^{K} w_{3k}} \sum_{k=1}^{K} w_{3k} T_{ik}.$$

The drift component will have a standard deviation of $0.01/\sqrt{ni(t)}$. However, in these simulations we only know about the first trait; the other trait still affects λ_i but we don't have any information about it.

I start with 20 species and two independently distributed traits whose interspecific standard deviations are both 30. The value of w_{31} is 0.01 and the value of w_{32} (the "missing" trait) varies from 0 to 0.05 to cover the range of cases from those in which the missing trait has relatively little effect on the average per capita immigration rate to cases in which it has 2.5 times stronger effects than the trait of which we are aware. For each case I did 100 independent simulation runs. The result is shown in Figure 5.13. As expected, the degree to which leaving out traits will affect the predictive ability depends on

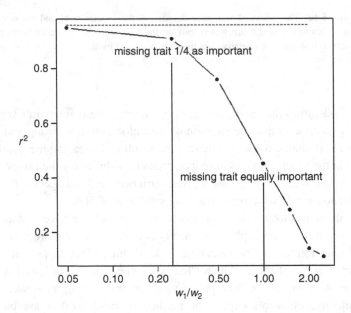

Figure 5.13. The effect of the importance of the missing trait (w_2) relative to the measured trait (w_1) in reducing the predictive ability of the Maximum Entropy Formalism, based on the bias-corrected r^2 value between observed and predicted values. The two weights (w_1, w_2) measure the effect of each trait on the average immigration rate of the species. The horizontal broken line is the predictive ability when the missing trait has no effect on immigration rates.

the relative importance of these missing traits in determining the demographic parameters. When the missing traits have little effect (or do not vary much) relative to the traits that have been measured then the decrease in predictive ability is modest. This is because the measured community-aggregated traits contain most of the relevant information about natural selection. When the missing traits do have big effects on the demographic parameters then the predictive ability decreases quickly. When the missing trait in Figure 5.13 is equally important relative to the trait that we have used for making the predictions, the bias-corrected r^2 value is reduced by just more than half. When the missing trait is 2.5 times more important than the trait that we have used for making predictions then the predictive ability is close to zero. These qualitative trends do not depend on the number of missing traits but only on their cumulative importance in determining the demographic parameters. Missing one important trait is worse than missing lots of relatively insignificant ones.

The quantitative results of these simulations should not be taken seriously since they depend on the details of the demographic model that is used. However, the qualitative conclusions should be robust to different demographic processes because they come from Equation 5.3. What are the practical implications of these results for predicting community assembly in real plant communities?

The first important conclusion is that we must think carefully about which traits will be most important in determining the demographic processes that affect community assembly. What traits affect the ability to disperse propagules over the landscape to various distances and the number of propagules produced? What traits affect the ability of propagules to survive and germinate, the ability of seedlings to survive and their subsequent growth rates? What traits affect the ability of different plants to survive until reproduction relative to competition, herbivory and abiotic stresses? What traits affect the timing of reproduction relative to disturbance regimes, and so on? Stated like this, our first conclusion is that we must know the functional ecology of the plants along the environmental gradients that we are studying and make sure that we obtain information on the most important traits.

The second conclusion points to a weakness of the current model. Throughout this book I have been talking about "species pools" without being very precise about what constitutes such a pool. The simulation results show that including lots of species in the "species pool" that are so poorly adapted to the site that they are almost never actually found will reduce the predictive ability of the model. Of course, including lots of species into the species pool that have little chance of immigrating due to non-trait factors is also a good way of reducing the predictive ability of the MaxEnt model, but we can deal with this, in principle, as described later in Chapter 8. However, the problem that we are dealing with here is rather different. Our problem here is to exclude species that are so poorly adapted to the conditions in our local community that they will never be found in the samples (because they will be exceedingly rare) even if they are common in the landscape. At this point I would like to give you a definitive solution but I don't have one. One (rather inelegant) way around the problem might be to first apply the Maximum Entropy Formalism to your large species pool and then calculate the predicted absolute abundance of each species based on the total abundance of all species at the site. Any species whose predicted relative abundance results in less than one individual could then be excluded and the analysis redone on the new reduced species pool.[27] We can only know if this will work by using real data.

---

[27] The species richness of a sample is the number of species in it having at least one individual each.

6

Community assembly during a Mediterranean succession

Theory can be dangerously seductive. Once one has built up an argument that is internally consistent, and once conclusions appear to follow inexorably from premises through clean lines of logic, it is sometimes enticing to conflate logical argument with reality. A good defense against such logical seduction is to let Nature into the conversation. A proper empirical evaluation of the method presented in Chapter 4 would involve an accurately measured environmental gradient involving all of the relevant environmental variables driving natural selection plus measured values of the key functional traits that respond to this selection of all species in the regional pool. This would be replicated in different localities along with evidence of quantitative generality of the community-aggregated traits. Hopefully, this book will have sufficiently convinced you of the potential of the approach that you will contribute to the hard work of assembling such empirical information. When this is done then we will know if the model actually works.

I'm easy to seduce. I think that it will work. However, I'm old enough to know the difference between seduction and commitment and I have had enough experience with field ecology to know that it might not work after all. I certainly won't hang myself in the barn if the model fails. As Thomas Henry Huxley famously pointed out, many beautiful theories have been killed by ugly facts.

At this stage[1] I can only offer one published empirical application of the MaxEnt model of trait-based environmental filtering (Shipley *et al.* 2006b). This study involved secondary succession in Mediterranean fields following abandonment of vineyards. The study site (Pic Saint Loup, 43°51′ N, 3°56′ E, 100 to 160 m asl) is just north of Montpellier (France) in an area that had

---

[1] I am aware of three other, unpublished manuscripts that have also found good predictive ability for the MaxEnt model. If reviewers are kind then these will be published in the near future.

suffered rather severe degradation of natural areas in earlier centuries but has experienced a partial recovery during the latter part of the twentieth century (Debussche *et al.* 1999). The empirical result of the test, as you will see, is very promising but the study was not originally designed to test the model at all and so the study has some important limitations that must be kept firmly in mind (especially *my* mind) in order not to exaggerate the degree of success from what is, in truth, only a preliminary test. These limitations will be discussed later. The climate is sub-humid Mediterranean with around 100 cm of precipitation each year, primarily from September to May with hot dry summers, and maximum/minimum average temperatures of 27 and 0 °C. At the end of the Second World War there began a process of abandonment of fields at this site that had previously been used as vineyards. Once agricultural activities ceased the natural vegetation began to re-establish following typical patterns of secondary succession in the area. None of these fields experienced subsequent major disturbances. The natural history of secondary succession in this area is given by Escarré *et al.* (1983).

A group of researchers at the Centre National de Recherche (Montpellier) have been studying the ecology of this successional sequence for many years. As part of this research, they identified 12 abandoned fields within a 4×4 km^2 area at the base of Pic Saint Loup that had all been vineyards before abandonment. They all have the same bedrock and the same soil type (brown calcareous or calcic cambisol: pH [H$_2$0] between 8.1 and 8.6), experience the same local climate, and all have similar histories of land use except for the year in which they were abandoned. The year of abandonment of these sites was determined using areal photographs, owner recollections and ageing of the dead rootstocks of the grape vines still in the soil. These ages varied from 2 to 42 years post-abandonment at the time of the study. The ecology of these sites has been described in a series of publications (Garnier *et al.* 2001, 2004, Kazakou *et al.* 2006, Shipley *et al.* 2006b, Vile *et al.* 2006a, b). The aboveground biomass of all species was collected at the peak of annual production in two 0.5×0.5 m^2 quadrats per site and separated to species; no woody species occurred in these sites. This data provided the estimates of relative abundance. Although over 100 species were recorded in these fields many of these were transient, occurring as single individuals in only one quadrat. We chose species that, together, accounted for at least 80% of the biomass in each site and these defined the species pool.

A number of typical patterns emerge (Escarré *et al.* 1983). In terms of community change, a set of annual species (*Erodium ciconium, Medicago lupulina, Medicago minima, Sanguisorba minor, Crepis foetida*) is most abundant immediately following abandonment but quickly decreases in abundance over time. Another group of species, dominated by *Picris hier-acioides*, increases in abundance at intermediate stages of secondary succession and then decreases. A third group of species, especially the grasses *Brachypodium phoenicoides* and *Bromus erectus*, increases in abundance over time and dominate the later herbaceous stage of secondary succession in these plots.

Denis Vile, during his PhD research (Vile 2005), set out to quantify the changes in the functional traits of the vegetation in these sites as a function of successional age. In essence, he wanted to see if one could model the relationships between functional traits and thus mathematically express the notion of a plant "strategy". To do this he measured 11 functional traits for each of 34 herbaceous species (18 annuals, 2 biennials and 14 perennials). Each species was also assigned to one of three stages of secondary succession in this area: early (0–6 years), intermediate (7–15 years) or the advanced herbaceous sere (15–45 years) based on when each species reached maximum abundance; this habitat descriptor for position along a successional trend is analogous to an Ellenberg Indicator Value (Ellenberg 1978). The trait covariation was modeled using structural equations and he successfully produced a path model showing how these functional traits are linked and change together (Vile *et al.* 2006b).

These 12 variables were highly correlated across species and a principal component analysis using standardized values revealed that the first three axes of variation accounted for 45%, 20% and 15% of the total variance respectively. The species' positions along the first two principal axes show that they form fairly homogeneous groups with respect to their occurrence of peak abundance during secondary succession (Figure 6.1). Species that occur together in time resemble each other, with respect to these functional traits, more than they resemble species occurring at a different time during secondary succession. The first axis shows a general trend for increasing plant size, later flowering and later seed maturation in the late successional species. The second axis contrasts annual species, characterized by high reproductive biomass and producing many small seeds, with perennial species having lower reproductive biomass and producing fewer larger seeds (Figure 6.2).

It is clear that these functional traits covary and that this covariation is somehow related to the typical successional position of each species, but how are these functional traits related together? Is it possible to produce a quantitative description of this covariation that reflects the likely causal relationships and tradeoffs of these traits? Using structural equations modeling, Vile produced a model that successfully accounts for all systematic covariation between the traits (Figure 6.3).

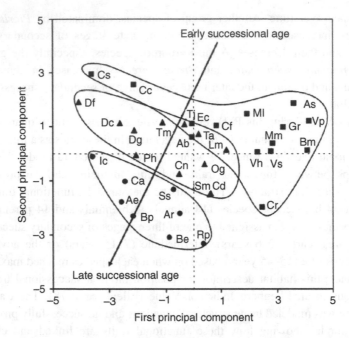

Figure 6.1. The position of each of 34 herbaceous species, occurring during a secondary succession in Pic Saint Loup, on the first two principal components. The ordination is based on 11 functional traits (see Figure 6.2), not species abundances. The circles group together species classified as occurring during early (squares), intermediate (triangles) or advanced seres during the secondary succession.

For more information on how to test the validity of such a structural equations model, see Shipley (2000a). To interpret Figure 6.3, note that there are two types of variables. Observed variables, enclosed in rectangles, are the 11 functional traits plus the habitat descriptor for the successional position of each species ("field age"). See Vile *et al.* (2006b) for the transformations used to obtain normality and linearity. The field age was coded as the mid-value of each time period: 3 for early successional species (years 0–6), 10 for mid-successional species (years 7–15) and 25 for later successional species (years 16–45). Variables inside ovals are latent variables that are not directly measured but whose value is inferred from the topological structure of the covariances of the other variables. The latent "successional index" therefore represents the correlated changes in those properties of the sites that are exerting selection pressures. The latent variable "plant size at seed maturation" is the total size of the plant at this phenological stage that induces allometric changes in the four measured components of plant size (height,

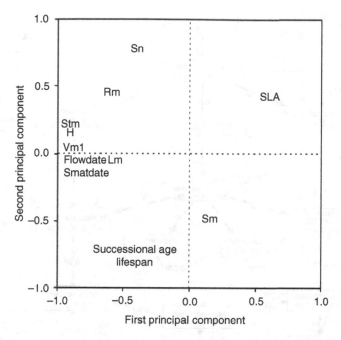

Figure 6.2. The position of 11 functional traits plus one environmental property (early, intermediate or late successional species) measured on 34 herbaceous species in the Pic Saint Loup secondary succession. Functional traits are: specific leaf area (SLA), seed mass (Sm), seed number (Sn), reproductive mass (Rm), stem mass (Stm), vegetative height (H), vegetative mass before flowering (Vm1), total leaf mass (Lm), Julian dates to flowering (Flowdate) and to seed maturation (Smatdate) and lifespan (annual/perennial).

plus the masses of the leaves, stem and reproductive tissues at the reproductive stage). The numbers along the arrows are path coefficients which are interpreted in the same way as a partial slope in a multiple regression. Note that, since "lifespan" was a binary (0 = annual, 1 = perennial) trait its path coefficient is the slope of a logistic regression. The numbers inside the rectangles (and ovals) give the intercept of the relationship and the proportion of the variance in the variable that is explained (R^2) by its direct causes. Finally, the number that is not enclosed by any form and whose arrow points to either an observed or latent variable gives the standard deviation of the residual variation of the variable, that is, the variation due to unknown independent causes that are not included in the model and that specifies by how much each species can deviate from the general "strategy". Because the resulting model, when translated into structural equations, successfully

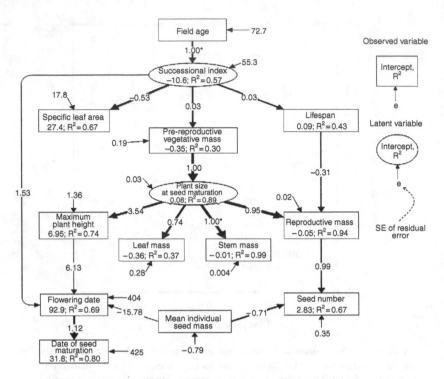

Figure 6.3. A structural equation model describing how the functional traits of the species in the Pic Saint Loup succession are linked. Variables inside boxes are the measured traits. Variables inside circles are unmeasured (latent) variables. Shown are the path coefficients, intercepts, error variances and r^2 values for each variable.

accounts for all patterns of covariation between the traits up to sampling fluctuations this means that the notion of an ecological "strategy", as described by Grime (1979, 2001), has been translated into a quantitative statement. If we can also quantify how the average trait values (i.e. the community-aggregated traits) change along the successional gradient, then we could use these equations to describe the multivariate selection gradient.

Figure 6.4 (which is the same as Figure 3.10) shows the community-aggregated values of eight of these functional traits. During secondary succession the overall vegetation changes from one dominated by annuals to one dominated by perennials, the average unit of biomass produces fewer seeds, the seeds mature later in the growing season, the specific leaf area of the average unit of biomass decreases, average vegetative mass, leaf mass and plant height increases while stem mass increases and then decreases. The lines in Figure 6.4

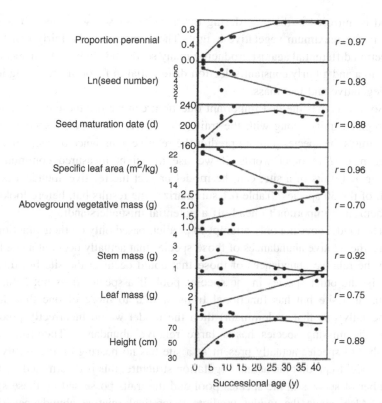

Figure 6.4. Observed and predicted relationships between eight community-aggregated traits and the successional age of each site (number of years post-abandonment). Lines show cubic-spline regressions.

are from cubic-spline smoother regressions. We see that the rate of change of the annual/perennial community-aggregated trait is rapid at first. Remembering the Breeder's Equation $(\Delta \bar{T}(t) = h^2(t)S(t))$, and knowing that the annual/perennial life history has a high heritability when comparing across such different genotypes, this means that, although annuals dominate immediately following abandonment of the vineyard, there is a strong selection differential against annuals during the early stages of secondary selection. The selection differential slows down over time and, after about 25 years post-abandonment, becomes essentially zero when no more annuals are to be found.[2] This

---

[2] Remember that the Breeder's Equation is a statistical statement. The selection differential is zero only because the trait ("being an perennial plant") no longer changes. This does not mean that an annual plant, introduced into this late successional site, would not experience any selective disadvantage.

trend is mirrored in the time during the growing season when seeds mature and in the maximum vegetative height. There appears to be fairly constant selection differential against producing many seeds and having a high specific leaf area and a fairly constant selection differential in favor of producing lots of vegetative and leaf mass.

So now comes the first important question: can we use these community-aggregated traits, along with the trait values of each of the 30 species that constitutes the species pool, to predict the relative abundance of each species at each site? In other words, if we use the eight measured community-aggregated traits of a site, can the model predict the relative abundances of each of the 30 species? Table 6.1 summarizes the results but, before looking at them, it is important to head off a potential misunderstanding.

The model does not only attempt to predict, based only on their functional traits, the relative abundances of those species that actually occur at a site but also the relative abundances of those that could occur at the site but don't; that is, the other species in the species pool. If a species does not actually occur at a site but has functional traits similar to those of one that does (especially one that is dominant) then the model would incorrectly predict that this missing species has a large relative abundance. Therefore, the number of species actually present at a site has no bearing on the ability of the model to predict the actual vegetation structure; this is determined by the number of species in the species pool and the traits possessed by these species. Also, since the model predicts theoretical relative abundances (i.e. Bayesian probabilities), and a Bayesian probability of exactly zero means that it is logically *impossible* for a species to occur (which would exclude it from the species pool), all predicted relative abundances must be greater than zero. Therefore, in order to be accurate, the model must predict a relative abundance close to zero for all those species that are missing from the site based only on their functional traits.

In these analyses, summarized in Table 6.1, I used a uniform prior during entropy maximization. This means that any differences in relative abundance at the landscape level are assumed to be irrelevant to community assembly in each local site. This is surely untrue but is probably a reasonable approximation since all of these sites are within a very small spatial area (4×4 km^2), sites having different successional ages are mixed together within this small landscape area, and the sites are within an agricultural matrix with people and machines (and probably seeds...) constantly moving around.

So how well does the model predict the observed relative abundances in each site if we use all eight functional traits and the corresponding

Table 6.1. *The proportion of the explained variance (R^2) in the observed relative abundances of each of the 30 species (the species pool) in each of the 12 sites using all 8 functional traits and also using only those traits that are significant in a backwards selection process (reduced model).*

Site	Successional age	All 8 traits in model			Reduced model			Traits
		R^2	R^2_{adj}	$F_{9,\,21}$	R^2	R^2_{adj}	$F_{v1,v2}$	
1	2	0.92	0.90	30.01	0.92	0.90	37.11	1, 2, 3, 5, 6, 7, 8
2	2	0.44	0.24	1.88	0.31	0.23	2.92	1, 3, 6, 7
3	7	0.82	0.75	10.70	0.82	0.77	14.92	1, 2, 3, 4, 5, 6
4	8	0.77	0.69	7.90	0.65	0.61	12.30	2, 3, 4
5	8	0.91	0.87	22.49	0.90	0.88	35.23	2, 3, 5, 6, 7
6	11	0.86	0.81	14.45	0.84	0.83	35.17	2, 5, 7
7	12	0.85	0.79	13.12	0.85	0.79	13.12	1, 2, 3, 4, 5, 6, 7, 8
8	26	0.99	0.99	256.93	0.96	0.95	115.29	1, 3, 4, 6
9	29	0.99	0.99	1870.0	0.98	0.98	356.26	3, 4, 6
10	35	0.99	0.99	3369.7	0.99	0.99	7596.4	3, 4, 5, 6
11	40	0.99	0.99	1617.2	0.99	0.99	1038.2	2, 6
12	42	0.99	0.99	1942.1	0.99	0.99	2960.67	2, 3, 4, 5, 6

The bias-adjusted value R^2_{adj} and the F-ratio testing the null hypothesis of independence between the observed and predicted values are also given. For the full model the degrees of freedom for the F-ratio are 9 and 21. For the reduced model the degrees of freedom are $v1$ = number of traits in the model, $v2 = 29$ – number of traits in the model. All are highly significant ($p \ll 0.001$) except for site 2 ($p = 0.19$ for full model and 0.04 for the reduced model). The functional traits are: lifespan (1, 0 = annual, 1 = perennial), vegetative mass before flowering (2), total stem mass (3), total leaf mass (4), vegetative height (5), Ln(seed number) (6), specific leaf area (7), Julian date at seed maturation (8).

community-aggregated values? With the exception of one site (site 2), the model predicts the actual relative abundances very accurately (Table 6.1). Sites 3 to 7, aged from 7 to 12 years post-abandonment, are predicted slightly less accurately but these are also the sites in which the community-aggregated trait values seem the most variable (Figure 6.4). If we compare the observed and predicted relative abundances of all 30 species over all 12 sites together, then the R^2 value is 0.94.

Can we obtain the same predictive ability with even less traits? This is a difficult statistical question for which I don't have a good answer. One obvious approach is to use a backwards step-wise procedure analogous to that in multiple regression. Since we can test for the significance of adding or subtracting variables in a nested series (Chapter 4) I therefore removed all those functional traits that did not significantly contribute to statistical improvement using the traditional significance level of 0.05 and this backwards step-wise procedure. Table 6.1 shows the result. Some traits could be removed without compromising the predictive ability in every site but one (site 7). In one site (site 11) only two traits were needed to give virtually perfect predictive ability. However, these results should not be interpreted in any causal sense. This statistical test suffers from the same problem as in the analogous multiple regression case because the traits are tightly correlated with each other. The high level of colinearity between the traits means that statistical tests of significance will be very sensitive to small (even random) fluctuations in the trait values and the resulting parameter values of the model (the λ values, i.e. the Lagrange multipliers, in this case) will be equally sensitive to such small changes.

If you go back to Chapter 4 you will recall that the equation that results from maximizing the entropy conditional on the constraints is $\ln(p_i) = -\lambda_0 - \Sigma_j \lambda_j T_{ij}$, where T_{ij} is the jth functional trait of species i. The values of T_{ij} don't change over environmental gradients since we are assuming (Chapter 5) that intraspecific variation can be ignored. The values of λ_j measure the importance of trait j in determining relative abundances[3] after controlling for the effects of the other traits in the model. However, the λ_j values will change to reflect the changing selection differentials on the trait in the different environments. Obviously, this means that the predicted relative abundances of the species (p_i) will also change over the gradient. In our Mediterranean succession the "gradient" is quantified by the post-abandonment age of the site. If we let the functions describing how λ_j *changes* over successional time (t) be $\lambda_j = f_j(t)$ and take the derivative with

---

[3] Actually, the relative abundances on a logarithmic scale.

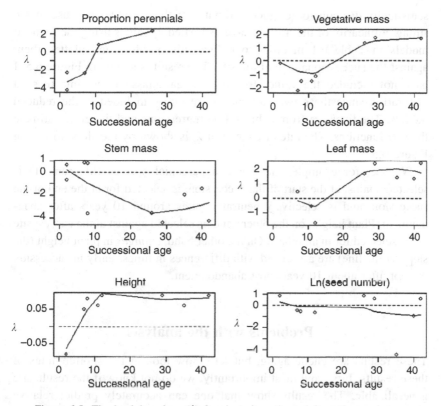

Figure 6.5. The lambda values (λ) for six traits estimated from the maximum entropy model estimated at each site, using the reduced models listed in Table 6.1. The lines are cubic-spline smoothing regressions.

respect to the successional time, then:

$$\ln(p_i) = \left(-\lambda_0 - \sum_j \lambda_j T_{ij}\right) = -f_0(t) - \sum_j f_j(t)$$

$$\frac{d(\ln(p_i))}{dt} = \frac{1}{p_i}\frac{dp_i}{dt} = \left(\frac{-d\lambda_0}{dt} - \sum_j \frac{d\lambda_j}{dt}\right) = -\frac{d(f_0(t))}{dt} - \sum_j \frac{d(f_j(t))}{dt}.$$

$$\text{(Eqn. 6.1)}$$

This equation links the rate of change of the relative abundance (and therefore of per capita growth rate and fitness, as shown in Chapter 5) with the rate of change of the lambda values. As explained above, it is statistically difficult to get good estimates of the λ_j values because of the strong colinearity of the traits. I have estimated λ_j values using the reduced models in Table 6.1 in order to reduce this problem and plotted them against the successional age of the site. The result is shown in Figure 6.5; I have not included the values for specific leaf mass or the time to seed maturation since these two variables were rarely included in the reduced models. The lines are from cubic-spline regressions and therefore estimate the $f_j(t)$ functions. The rate of change of λ_j is shown by the slope of line in Figure 6.5.

We see, for example, that being a perennial species (lifespan = 1) is selected against at the start of the succession, is selected for at the end of the succession, and is selectively neutral at sites around 10 years after abandonment. Plant height, on the other hand, is selected against at the earliest site but is selected for in all others. On the other hand, *changes* in plant height (the slope of the line) are associated with differences in fitness early in succession but not after about 10 years post-abandonment.

Problems with the analysis

These results are encouraging, but let's now look at the shortcomings of these results. First, and most importantly, we do not know if the results are generalizable. The results show that one can accurately predict relative abundance if we know the community-aggregated trait values at a site but this, alone, is not really very useful as a practical matter. After all, we had to measure the actual relative abundances at each site in order to get the community-aggregated traits! This exercise is not circular, as explained in Chapter 4; it is the same as regressing Y on X in some set of data and then testing for significance and quantifying the strength of the relationship. In such simple regression analyses we first use the data to estimate parameters (the slope and intercept) and this uses up two degrees of freedom. We then use the remaining (residual) degrees of freedom to test and quantify the relationship. However, true predictive ability requires generality and this means that we should get new sites whose successional age is known, predict the community-aggregated traits in these new sites given their successional ages, and then predict the relative abundances of the species. Since I don't have any new sites, I don't know if it will work. In Shipley *et al.* (2006b) we did redo the analysis using the predicted community-aggregated trait values (the lines in Figure 6.4) rather than the observed ones, and this resulted in good predictive ability, but this is to be expected since the observed and predicted community-aggregated traits were very close.

Marks and Muller-Landau (2007) have suggested that one use cross-validation techniques to evaluate the predictive ability of the model. This involves leaving one site out of the data set, predicting its community-aggregated trait values using the general regression obtained from the remaining sites, and then predicting the relative abundances. This is a good idea in general, and I report the result of such a cross-validation in Shipley (2007) but it requires a rather larger number of sites than are available (another weakness). This is especially important when the relationships between the community-aggregated traits and the environmental variables are strongly nonlinear (as is the case with these data) since leaving out a single site that is located at a point along the gradient where the community-aggregated trait value is changing rapidly will result in a large prediction error.

Actually, I would not expect the results to be generalizable beyond (perhaps) the same local area in southern France even if I had a much larger number of sites, and this leads to another weakness with the analysis. The environmental variable that was chosen to quantify the gradient was "years post-abandonment". This is a poor choice[4] for our purposes. "Time" doesn't cause vegetation to change during succession. The causally efficacious variables are those environmental variables that change *in* time and these environmental variables can be expected to change more or less quickly in different areas, thus frustrating any general relationships between community-aggregated traits and time following abandonment.

What other limitations exist in this analysis? Well, the species pool (30 species) is not very large. Ideally we would want a larger spatial scale than just 4×4 km^2 and a much larger species pool. Given the large trait data bases that are now being assembled, it should soon be possible to have species pools of several hundred species. Of course, such analyses would be limited to those traits that where chosen for inclusion and these would not necessarily be those that are being most strongly selected by the environmental gradients that exist in the actual vegetation. Similarly, the measured environmental variables might simply be those that are easiest to measure rather than being those that are guiding natural selection.

A final weakness is that we have implicitly assumed that dispersal limitation from the larger landscape is irrelevant; this is because I used a uniform

---

[4] Remember that the study was not originally designed to test this model. The use of successional time was perfectly appropriate for the original purpose of quantifying plant strategies during secondary succession.

prior in the analysis. As the size of the species pool (and the spatial extent from which these species come) increases, it is increasingly likely that causes other than traits become important in determining which species from the large species pool will occur in the local community, and at what relative abundance. Such non-trait-based effects could be incorporated by using a prior distribution that reflects the relative abundance of each species in the larger landscape, but such information is lacking. This is discussed further in Chapter 8.

7

The statistical mechanics of species abundance distributions

So far in this book we have been looking for an answer to the following general question: If we know which species exist in the species pool, and we know their functional traits, can we predict the relative abundance of each species in different environmental contexts? What happens if we don't know which species – or even how many species – are in the species pool? Clearly, if we don't know which species are in the pool then we are in no position to predict their relative abundances. Even if we do know that a particular species is in the pool we still can't predict its relative abundance if we don't know how many other species are in the pool; after all, relative abundance is a proportion from a total. Therefore, if we don't know the composition of the species pool then we can't possibly answer the central question posed in this book.[1]

However, even when missing this vital information about the composition of the species pool, we can ask a different, but related, question. Since we can't inquire about the abundance of species i, we might want to know how *many* species will have a given abundance. We can ask, for example: how *many* species will have only one individual or unit of biomass? How *many* species will be a little less rare and have two individuals, or units of biomass, and so on? The distribution of species having different levels of abundance in a community, irrespective of whether or not we know which species these are, is called the "species abundance distribution", or "SAD", and the prediction and explanation of such species abundance distributions has been a goal of community ecologists for a very long time.

This second question (predicting and explaining the species abundance distribution) is different from the main question posed in this book. I had two

---

[1] Actually this sentence is rather more negative than it should be. In a practical sense we must construct a species pool that includes those species likely to have appreciable abundance. Leaving out species that will always be rare will have no practical effect since the prediction is always a *relative* abundance.

reasons to exclude it (and therefore this chapter). First, the relationship between abundance and functional traits in this chapter will be tenuous at best, although I will sketch out a possible relationship later. This chapter will deal mostly with demography. Second, the modeling of species abundance distributions is already a growth industry despite its long history and a proper treatment of the topic would take an entire book to do properly. Despite this, I had one good reason that, for me, trumped the others. Pueyo *et al.* (2007), Dewar and Porté (2008) and Beninca *et al.* (2008) have shown how novel theoretical applications of the Maximum Entropy Formalism to the question of species abundance distributions can lead to a very different conception of the causes and origins of species abundance distributions. I have already described (Shipley *et al.* 2006b) the model developed in Chapters 4 and 5 as a "statistical mechanistic" approach to community assembly. In one sense this is justified since the approach consists of predicting macroscopic properties of communities by maximizing entropy subject to constraints. However, the constraints used in classical statistical mechanics consist of basic physical limits on totals (energy, mass, volume) while my model consists of constraints generated by natural selection. The theoretical results of Pueyo *et al.* (2007) and Dewar and Porté (2008) are much closer to the spirit of classical statistical mechanics since the constraints are those generated by physical limits of space, resources and population dynamics. In order to avoid some important misunderstandings (Haegeman and Loreau 2008) I should also explain that there are two, rather different, schools of "statistical mechanics". Classical statistical mechanics is based on a combinatorial argument and on a fundamental assumption that all microstates are equally likely. The Maximum Entropy approach, developed most completely by Jaynes and explained in Chapter 4, rests on very different assumptions and is essentially a process of logical inference (Shipley 2009). The results in this chapter can only properly be described as "statistical mechanistic" with reference to this latter school.

If the results of Pueyo *et al.* (2007) and Dewar and Porté (2008) prove to be correct then they will greatly simplify, both conceptually and mathematically, this important area of community and macroecology. Despite the fact that these two groups worked completely independently of each other, I received the manuscript versions of these two papers from two different journals for review and within a few weeks of each other. Given that the Maximum Entropy Formalism is not exactly a standard technique in ecology, the conjunction of these unlikely events suggests that something important is in the air. I caution you that the ideas are still mostly hypothetical but they could prove to be fertile soil for future empirical studies. This caution is especially important because the ideas are so intellectually elegant.

What is a species abundance distribution?

The questions posed in this chapter are slightly different from those posed in the rest of this book. Questions that are similar to one another, but are not identical, cause particular problems in understanding. It is important to clearly see the difference. Figure 7.1 (top, left) shows the hypothetical distribution of 98 individuals into 26 species (species A to Z) at a site. We see that only one species (A) had 30 individuals, that two species (B and C) had 10 individuals each, and so on. The top right graph of Figure 7.1 shows this same information expressed as relative abundances; i.e. the proportion of the total number of individuals at the site found in each of the species. This second graph, showing the relative abundance of each species in the species

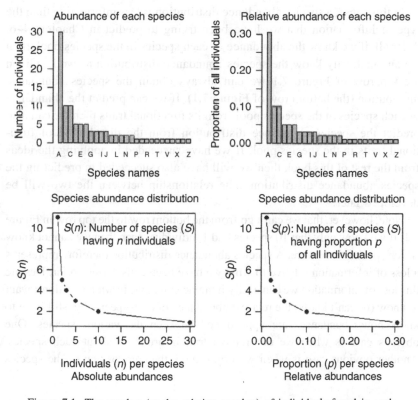

Figure 7.1. The number (or the relative number) of individuals found in each species in the species pool are shown in the top row. The number of species possessing different numbers (or relative numbers) of individuals are shown in the second row. The second row therefore gives the species abundance distribution.

pool, shows the type of information that Chapters 4–6 have attempted to predict and explain. We already know (in theory) how we should go about predicting and explaining such patterns.

The bottom row of Figure 7.1 shows the species abundance distribution. This is a type of information that we have not yet faced. The bottom left graph is simply a summary of the information in the top left graph with the identity of the species left out. It says, for example, that there are 11 species having only 1 individual. Notice that there is no way to know the identity of these 11 rare species. It is true that we know, from the top left figure, that this corresponds to species P to Z, but such information is missing from the species abundance distribution. A species abundance distribution simply tells us how many species (S) have an abundance of n individuals in the community (thus $S(n)$) without saying *which* species these are or how many individuals are found in any *given* species.

In this sense, a species abundance distribution is less informative than the type of information that we have been trying to predict in Chapters 4–6. After all, if we know the abundance of each species in the species pool, then we automatically know the species abundance distribution as well.[2] Given the top row of Figure 7.1 we can always obtain the species abundance distribution (the bottom row of Figure 7.1). If we can predict the abundance of each species in the species pool from its functional traits then we can also predict the species abundance distribution from the distribution of functional traits in the species pool. If we have succeeded in applying the ideas from the rest of this book then we will have also succeeded in predicting the species abundance distribution. The relationship between the two will be developed later.

Note, however, that we can't go from the bottom row to the top row in Figure 7.1. If we know only that 10 species had 1 individual each then we cannot know which species these were. A species abundance distribution therefore represents a loss of information relative to what we have been considering so far. Despite this loss of information we can easily imagine situations in which we don't want to know (or can't know) the relative abundance of each species but still want to know how abundances are apportioned between the various species. One obvious case is when we don't have trait information about each species. Another obvious case is when we don't know the composition of the species

---

[2] Of course, species in the species pool that are predicted to have very low relative abundance may well be missing completely from the actual community, especially if the spatial scale is small, simply because there is a fixed number of individuals and one cannot observe a fraction of an individual.

pool. As you will see, even without information about traits or species pools, we can still use the Maximum Entropy Formalism to predict the species abundance distribution. Indeed, the application of MaxEnt to such a problem is the closest analogy that we will find to a statistical mechanics of ecological communities. Before we do this, we should first take a detour into history because the prediction and explanation of the species abundance distribution is one of the oldest goals of community ecology, and one that is still actively pursued today.

A brief history of the species abundance distribution

The first mathematical equation that was proposed for a species abundance distribution (Equation 7.1) was published in Japanese by a Japanese researcher (Motomura 1932), as cited by Whittaker (1965). Ten years later Corbet, a British entomologist, proposed the same equation to predict the number of species (S) of Malaysian butterflies having different numbers (n) of individuals (Corbet 1942). Corbet was presumably[3] not aware of Motomura's work. The equation is:

$$S(n) = \frac{C}{n}.$$ (Eqn. 7.1)

Corbet found that Equation 7.1 worked well in predicting $S(n)$ in all but the most common species, not only in Malaya but also for butterflies from the Tioman and Mentawi Islands. Equation 7.1 is also called the "geometric series" where C is an empirical constant that must be estimated from the data and so this equation simply claims that the number of species having n individuals is inversely proportional to n. In a subsequent publication Corbet teamed up with C. B. Williams and R. A. Fisher (Fisher *et al.* 1943) to propose a more general equation which they applied both to Malayan butterflies and to English moths. This equation, since baptized the "logseries distribution", has the form of Equation 7.2a but I will also express it in a different, but completely equivalent, form (Equation 7.2b) in which the constant "x" is replaced by the exponent of another constant $\bar{\omega}$ (i.e. $e^{\bar{\omega}}$):

$$S(n) = \frac{a}{n} x^n,$$ (Eqn. 7.2a)

---

[3] My only evidence for this is the fact that Corbet didn't cite Motomura in his paper. Given that Motomura's paper was written in Japanese, it is reasonable to assume that Corbet – an Englishman – didn't read it.

$$S(n) = \frac{a}{n}x^n = \frac{a}{n}(e^{\bar{\omega}})^n = \frac{a}{n}e^{\bar{\omega}n}. \qquad \text{(Eqn. 7.2b)}$$

In Equation 7.2a, b there are two empirical constants that must be estimated from the data (x or ϖ and a).[4] The logseries distribution was derived by Fisher from purely statistical assumptions. If the true relative abundance of a given species (say, species i) in the community is $\hat{r}_i$ and the true number of individuals of all species is N, then the expected number of individuals belonging to species i (i.e. the true abundance) is $N\hat{r}_i$. Fisher then made the assumption that the chances of a researcher observing different numbers of individuals of species i in a random sample are independent of one another. In other words, if you see an individual of species i then this doesn't change the chances of you seeing another individual of the same species. This is actually a rather unlikely assumption since individuals of a species usually tend to be clumped in space and time (especially for plants) but he made this assumption because, given it, he knew that the probability of observing n_i individuals of species i in a random sample follows a Poisson distribution if the total number of individuals (N) in the community are very large (in theory, infinitely large):

$$p(n_i) = \frac{(N\hat{r}_i)^{n_i}e^{(-N\hat{r}_i)}}{n_i!} = C\frac{(\hat{r}_i)^{n_i}e^{-\hat{r}_i}}{n_i!}.$$

Next, he reasoned that this probability distribution will hold for every species in the community except that the true relative abundance ($\hat{r}_i$) will differ for each species. If we knew how these true relative abundances were distributed amongst the species then we could proceed. Note that this is close to the problem that we have been tackling in Chapters 4–6, where we use the plant traits to predict these theoretical relative abundances. However, Fisher proceeded in a different way. Here are his words (Fisher *et al.* 1943, page 54): "*An important extension of the Poisson series* [i.e. the Poisson distribution] *is provided by the supposition that the values of m* [i.e. the true abundances, $N\hat{r}_i$] *are distributed in a known and simple manner. Since m must be positive, the simplest supposition as to its distribution is that it has the Eulerian form* [i.e. a gamma distribution] . . .". In other words, he simply assumed that the true abundances of each species follow a gamma distribution because it is the "simplest" assumption. Upon combining these two assumptions (a Poisson distribution of observing n_i individuals of species i in a sample, and a gamma distribution of the true abundances of each species in the community) he

---

[4] These two constants can also be calculated by solving the equations: $a^{-1} = \ln\left(\frac{\bar{a}}{a} - 1\right)$ and $\varpi = \ln\left(\frac{a}{\bar{n}} + 1\right)$.

obtained the distribution $S(n)$ of observing S species having n individuals each, which was a negative binomial distribution. Since very rare species will likely have zero individuals in the sample due to sampling fluctuations, we must exclude the case of $S(0)$. The final step in the derivation of the log-series distribution was to point out that all actual communities tend to have a few very common species and many rare species, meaning that the parameter in the negative binomial distribution controlling the heterogeneity of the true abundances will always be close to zero. From this, the logseries distribution (Equation 7.2) was born. It is important to keep in mind that the logseries distribution, like the geometric series distribution, is a purely empirical equation whose only virtue can be predictive ability. Remembering what it means for a theory to claim explanatory ability (Chapter 2), the logseries as derived by Fisher cannot possibly claim to "explain" data since its only assumptions (mutual independence of individuals in the community at both the intraspecific and interspecific levels, an infinite number of individuals in the community, and a distribution of true abundances per species following a gamma distribution) are either wrong or without causal content.

In 1948, F. W. Preston published another empirical equation for the species abundance distribution (Preston 1948). Working with birds, he proceeded as follows: rather than counting the number of species having exactly n individuals, he grouped together ("binned") all those species having individuals within each logarithmic unit. For instance, working with base-10 logarithms, one would count the number of species having between 1 and 10 individuals (i.e. between $\log_{10}(1) = 0$ and $\log_{10}(10) = 1$), then the number of species having between 11 and 100 individuals (i.e. between $\log_{10}(10) = 1$ and $\log_{10}(100) = 2$), and so on. One can use any other logarithmic base as well since they differ only by a constant; in fact, Preston used base-2 logarithms (thus bins of 1–2, 2–4, 4–8, . . .). After binning the data in this way, he noticed that the resulting pattern was closely approximated by a normal curve. Because the curve is normal using logarithmically transformed data, the curve is called the lognormal species abundance distribution (Equation 7.3). In point of fact, there is no good reason why one must bin the observations and one can more easily work directly with the log-transformed values themselves. If working with binned data then C_1 would be the number of species in the modal bin and C_3 would be related to the variance of the curve ($1/2\sigma^2$):

$$S(n) = C_1 e^{\frac{\left(\ln(n) - \overline{\ln(n)}\right)^2}{C_3}}.$$

(Eqn. 7.3)

Preston also pointed out that the lognormal distribution emerges graphically from the logseries distribution as the size of the sample taken from the community increases. This is because the rarest species will be missed in small sample sizes and you can not observe less than one individual per species. For instance, assuming a Poisson distribution for observing n_i individuals, if the true relative abundance of species i is 0.001 then, if you sample 100 individuals then you would have a probability of only about 0.1 of observing at least one individual of species i. On the other hand, if you sample 1000 individuals then you are reasonably likely ($p = 0.63$) to observe at least one individual of species i and with a sample size of 10 000 individuals then you are almost certain to find at least one individual of species i ($p = 0.99995$). Preston called this a "veil line" in which, as sample size increases, more and more of the full lognormal distribution is revealed (Figure 7.2). The lognormal distribution has been more successful than the logseries distribution at fitting empirical data (Sugihara 1980) although using binned data to empirically test curves is not a good idea since virtually any curve with a long right tail will tend to be approximately normal on a log scale.

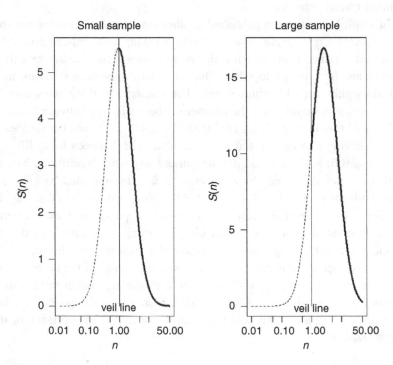

Figure 7.2. A graphical illustration of Preston's "veil line" for the lognormal species abundance distribution.

Preston later[5] described the "canonical" lognormal distribution (Preston 1962). If Equation 7.3 gives the number of species S having n individuals, then the product "$nS(n)$" gives the total number of individuals found in these $S(n)$ species. Preston called the function $f'(n) = nS(n)$ the "individuals" curve as opposed to the "species" curve, i.e. $f(n) = S(n)$. Since n can also be written $e^{\ln(n)}$, Preston pointed out that the function $f'(n) = e^{\ln(n)}S(n)$ is also a lognormal function of n with the same variance (σ^2) but with a mean that is shifted to the right of $S(n)$ by an amount equal to this variance. Now, if the number of individuals in the most abundant species is n_{max} then, for this single most abundant species (which we will call species d), $S(n_{max}) = 1$. Preston then made the assumption that this single most abundant species will have more individuals than any other group of subdominant species having equal abundances. In other words, $n_{max}S(n_{max}) > n_jS(n_j)$ for all $j \neq d$. If this is the case then the mean of $f'(n)$, which is $\mu + \sigma^2$, occurs at n_{max}. He also pointed out that the value of σ^2 did not vary much in empirical data sets and, assuming a constant value, this induced an allometric relationship between the total number of species in the community and the number of species having the modal number of individuals; i.e. $S\left(\overline{\ln(n)}\right)$.

Sugihara (1980) looked at many empirical data sets and showed, not only that the lognormal species abundance distribution fit such data sets very well, but that these empirical results agreed with Preston's canonical version. This was seen as having special biological importance for the following reason. Imagine arranging all of the N individuals (•) in the community in a line: • • •. . .• • •. Now randomly separate these N individuals into two groups:• • |•. . .• • •. Now randomly chose one of the two groups and repeat this same operation: • • |•. . .• • |•, and so on. Sugihara showed that if one were to continue doing this then the resulting distribution of individuals into groups (i.e. species) follows a canonical lognormal distribution. This explanation for the origin of the lognormal species abundance distribution has been criticized because there is no convincing biological mechanism of how such a random partitioning of individuals into species can occur and because there is no "stopping rule" to say when the division should end; that is, what determines the total number of species. After all, if the division goes on long enough then every individual will eventually be a different species. The second criticism is important but keep in mind the idea of individuals being randomly assigned to species. This is because we have already encountered the idea (Chapter 4) of individuals being randomly assigned to species (now subject to

---

[5] Interestingly, Preston chose the word "canonical" by analogy with the term used by J. Willard Gibbs in statistical mechanics.

constraints) but with a biological explanation (Chapter 5). However, the implied biological importance of the canonical version of the lognormal curve is unclear. As Limpert *et al.* (2001) have pointed out, the lognormal distribution occurs widely in nature in many non-ecological contexts (concentrations of elements in rocks, latency periods of diseases, rainfall amounts and even the size of oil drops in mayonnaise) and the variance of the lognormal distribution for a given phenomenon is usually quite constant. In fact, based on Table 2 of Limpert *et al.* (2001), the range of the variance of the lognormal species abundance distribution is actually wider than for many other non-ecological examples. Nekola and Brown (2007) have recently pointed out many other non-ecological examples and argue that the lognormal distribution is due to the complex dynamics of such systems. We will look at this more formally later in this chapter.

There have been quite a few other attempts to derive models for the species abundance distribution, involving various assumptions, mechanisms and mathematical methods. McGill *et al.* (2007) provide a good review. However, given its recent prominence in the literature, I want to look at one more.

Neutral models and the species abundance distribution

I spent much of Chapter 2 trying to explain why demographic models of community assembly are doomed to fail by the criteria of this book. The inherent nonlinearity of functions describing changes in per capita growth rates, the feedback relationships between the species, and the interactions with changing environmental conditions result in models so complex, and so potentially sensitive to small parameter changes, that we will not be able to apply them to real populations in the field. What if we were to remove all this complexity by fiat? What if we were to simply *assume* that the probabilities of birth, death and immigration were constant and the same for all individuals of all species? Since the probabilities of birth and death of a genotype together define its absolute fitness, our question reduces to assuming equal absolute fitness values for all species.[6] If we let this equal absolute fitness value be f then, remembering the relationship between absolute fitness and per capita growth rate (Chapter 5), we can write $\frac{1}{n_i(t)}\frac{dn_i(t)}{dt} = f + \varepsilon_i(t)$ for all species, where f is a constant and $\varepsilon_i(t)$ is a random component drawn from the same

---

[6] Since the absolute fitness values of all species are assumed to be equal, and necessarily greater than zero or else there would be only bare ground, then the relative fitness values of all species are equal to zero.

sampling distribution, and arises from sampling fluctuations in a finite population. In fact, we can go even further. Since the absolute fitness (f) is the same for all species then the per capita growth rates must also be equal for all species except for random fluctuations around this equal value. But this constant f can only be either zero, less than zero or greater than zero. If f is less than zero and is the same for all species then the populations of all species are exponentially decreasing (except for random fluctuations) and this means that the site will eventually be denuded of all plants. Clearly, f cannot be less than zero. If f is greater than zero and equal for all species then the populations of all species are exponentially increasing without anything to slow them down. Clearly, f cannot be greater than zero either in a world of finite resources.[7] Therefore, by assuming that f is equal for all species, we are effectively assuming that f is *zero* and so we end up with $\frac{1}{n_i(t)}\frac{dn_i(t)}{dt} = \varepsilon_i(t)$ for all species. This is a "neutral" model and, because of the extreme simplicity of this system of equations, it has proven possible to derive analytic formulae for the expected species abundance distribution at equilibrium.

As we saw in Chapter 5, a neutral model in which no new species are entering through immigration will eventually result in all but one species being driven to local extinction through random drift. This would not be a very useful species abundance distribution! However, things get more interesting when the model is divided into two parts: a "metacommunity" that exists at a large spatial scale and that defines the species pool,[8] and a "local community" that exists at a smaller scale and into which new individuals are constantly being introduced by immigration. Of course, to remain neutral the model must assume that the per capita probability of immigration is also equal for all species. Since the actual per capita growth rates of the species are determined only by random sampling fluctuations, this means that the process of immigration into the local community is also determined only by random sampling fluctuations and forms part of $\varepsilon_i(t)$.

Although the first explorations of such a neutral process relied on computer simulations, Volkov *et al.* (2003) and Etienne (2005, 2007) have derived analytical formulae for this process. Here, I present the derivation of Volkov *et al.* (2003). We start in the metacommunity; that is, the collection of individuals and species at some larger spatial scale that excludes those of the local community, but whose individuals can potentially immigrate into the local community. We will write "$b_{n,k}$ and $d_{n,k}$" to mean the probability of

---

[7] Including space!

[8] A metacommunity is essentially what we have been calling a "species pool" together with a description of the relative abundance of each species at this larger spatial scale.

species k, having n existing individuals, giving birth (b) to, or losing through death (d), individuals. The symbol "$b_{0,k}$" represents the addition (i.e. "birth") of an individual that does not yet exist in the metacommunity; this could represent a speciation event or the immigration of a new species from some even larger metacommunity. Since the model is neutral, meaning that the per capita probabilities of birth and death (thus fitness) are equal, we will later forget about the species subscript (k). Volkov *et al.* (2003) show that the steady-state or equilibrium solution to such a stochastic process is Equation 7.4 where version 7.4a is for the case where species do not necessarily have equivalent fitness and where 7.4b is for the special case of neutrality. $P_{n,k}$ and P_n are the probabilities of species k (or all species) having n individuals and $P_{o,k}$ *and* P_o are normalization constants to insure the probabilities sum to unity:

$$P_{n,k} = P_{o,k} \prod_{i=0}^{n-1} \frac{b_{i,k}}{d_{i+1,k}}$$

$$P_n = P_o \prod_{i=0}^{n-1} \frac{b_i}{d_{i+1}}. \qquad \text{(Eqn. 7.4a, b)}$$

It is important to point out that this derivation does not assume that the total number of individuals in the metacommunity are fixed over time, unlike Hubbell's "zero-sum multinomial" (2001), but what does it assume? Although not explicitly stated in the original publication, the assumption of a "steady-state equilibrium" means that the metacommunity converges to a constant total number of individuals $(\hat{N}_M)$ and a constant total number of species $(\hat{S}_M)$ and this, in turn, means that the average number of individuals per species in the metacommunity, $\bar{n}_M = \hat{N}_M / \hat{S}_M$ becomes constant; I have included the subscripts (M) to remind you that these are values in the metacommunity. This same assumption would also hold if the total number of individuals in the metacommunity was fixed. This point will be important later when we look at a MaxEnt solution. When we combine Equation 7.4b for all species to get the predicted number of species in the metacommunity having n individuals, $S_M(n)$, we get Equation 7.5:

$$S_M(n) = S_M P_o \frac{b_0 b_1 \dots b_{n-1}}{d_1 d_2 \dots d_n}$$

$$S_M(n) = S_M P_o \frac{b_0 (1b) \cdots ((n-1)b)}{(1d)(2d) \cdots (nd)} = S_M P_o b_0 \frac{1}{n} \left(\frac{b}{d}\right)^n$$

$$S_M(n) = \theta \frac{x^n}{n}$$

$$S_M(n) = \theta \frac{(e\lambda)^n}{n} = \theta \frac{e^{\lambda n}}{n}. \qquad \text{(Eqn. 7.5)}$$

The neutral model for the metacommunity predicts the Fisher logseries distribution![9] As an aside, this metacommunity absolutely requires the input of new species, either through speciation or through immigration from some other metacommunity, or else the equilibrium metacommunity would consist of a single species, like all processes of random drift.

Now, given this metacommunity, what happens in a local neutral community that follows the same rules, with new immigrants arriving from the metacommunity? The important thing to remember is that this local community will have many fewer individuals that does the metacommunity and so random drift will be more pronounced. Volkov *et al.* (2003) derive the result, as given in the intimidating[10] Equation 7.6. In this equation we must know (or estimate from empirical data) three things: the immigration rate of new individuals from the metacommunity (m), the total number of individuals in the local community (N) and a numerical constant (θ):

$$S_L(n) = \theta \frac{N!}{n!(N-n)!} \frac{\Gamma(\gamma)}{\Gamma(N+\gamma)} \int_{y=0}^{y=\gamma} \frac{\Gamma(n+y)}{\Gamma(1+y)} \frac{\Gamma(N-n+\gamma-y)}{\Gamma(\gamma-y)} e^{\frac{-y\theta}{\gamma}} dy$$

$$\gamma = \frac{m(N-1)}{1-m}$$

$$\Gamma(x) = \int_{y=0}^{y=\infty} y^{x-1} e^{-y} dy. \qquad \text{(Eqn. 7.6)}$$

The form of Equation 7.6 approaches a logseries distribution when the immigration rate (m) becomes large; essentially, the local community is entirely determined by the structure of the metacommunity. If the immigration rate becomes zero then the local community is entirely isolated from the metacommunity and the local community would eventually be reduced to a single species through random drift (Chapter 5). When the immigration rate is low and N is not too large then Equation 7.6 is like a logseries distribution except for a lower number of very rare species. Why does the local community show patterns that are different from those in the metacommunity? It is because, although the *per capita* immigration rate is the same for all species (i.e. the assumption of neutrality at the level of individuals), the immigration rate per species is not the same for all species. The immigration rate per species depends both on the per capita immigration rate (which is the same for all species since each individual has the same probability of immigrating

---

[9] The a and ω constants in Equation 7.2b are the θ and λ constants in Equation 7.5.
[10] The equations of Etienne are equally intimidating.

into the local community) and on the actual number of individuals per species in the metacommunity. Those species that are most abundant (by pure chance or due to other reasons) in the metacommunity will contribute the most immigrants to the local community.

In judging neutral models we might concentrate on predictive ability, on explanatory ability, or perhaps on simplicity of use. Zillio and Condit (2007) have recently claimed that the value of neutral models lies "... *in their simplicity, in their ability to provide a manageable way to calculate some properties of the system, and as a starting point to develop more refined theories*". A look at Equation 7.6 should convince you that such models are not mathematically or computationally simple. In fact even estimating the integral in Equation 7.6 is numerically challenging[11] and fitting the logseries or the lognormal distributions are very much easier to do. Perhaps Zillio and Condit are referring to explanatory simplicity although it is not clear from their words? We will evaluate explanatory ability later.

How about predictive ability? McGill *et al.* (2006) conducted an extensive meta-analysis of all known empirical studies using neutral models and compared the predictive ability of Equation 7.6 versus the simpler lognormal distribution based on nine different measures of goodness of fit. In fact the neutral model performed worse than the lognormal on eight out of nine measures.

Finally, what about explanatory ability? It is certainly true that mathematical models like Fisher's logseries or Preston's lognormal do not have explanatory ability in any biological sense. Fisher derived his logseries simply from statistical assumptions about sampling and Preston obtained his lognormal in a purely inductive manner by choosing a convenient mathematical function (the lognormal) that provided a good description to empirical data. As explained in Chapter 2, for a model to claim explanatory ability one must show both that the assumptions agree with empirical evidence and that the predicted patterns also agree with empirical evidence. Now, if you are an ecologist rather than a mathematician (and you are still reading this book after encountering Equation 7.6) then you are surely asking why one would make such an absurd assumption as neutrality. After all, "neutrality" is another way of saying that all genotypes of all species have exactly equal fitness values. If we could show that every genotype of every species really did have equal fitness then we will have made Creationists very happy. In particular, this would mean that evolution by natural selection was impossible, it would annul over a hundred years of careful field work

---

[11] Estimating the parameters in large data sets can take hours on a fast computer.

documenting fitness differences between genotypes, and it would exclude the possibility of changes in community-aggregated traits over environmental gradients (Chapter 5). In fact, if this assumption was correct then deriving the SAD would be easy; if all genotypes truly had equal fitness then evolution by natural selection would be impossible, we would have no ecological communities whose abundance distribution we need to explain, and no ecologists interested in the answer! However, some very good ecologists have made just such an assumption in full knowledge of its implications if taken to its logical conclusion. They have done this in order to define a sort of "null" model for the species abundance distribution. As with all null models, its importance is only in quantifying by how much reality deviates from it. It would, to again cite Zillio and Condit (2007), provide a *"starting point to develop more refined theories"*. If the resulting neutral species abundance distribution was contradicted by the known empirical patterns (McGill *et al.* 2006) then this would provide evidence for niche-based explanations that require fitness differences and would point the way to such "more refined" theories. However – and this point has been largely missing in the large literature debating neutral models – the converse is *not* true. If the empirical patterns can be generated by such a neutral model then it does not logically follow that this agreement provides support for neutral assumptions. The *IF–THEN* structure of the logical argument can only go forwards when claiming explanatory ability, never backwards (Chapter 2). Assumptions imply conclusions but conclusions do not imply assumptions.

Maximum entropy

This next section takes a completely different approach to the species abundance distribution, namely a statistical mechanistic approach[12] based on the maximum (relative) entropy formalism. As we have seen, previous attempts have relied on one of two strategies: either simply looking for a function that best describes empirical patterns based on goodness of fit (like the logseries or the lognormal) or use demographic models that, of necessity, must make unreasonable assumptions in order to solve the mathematics (like neutral models). A statistical mechanistic approach instead asks what sorts of general constraints might exist during community assembly from an unknown

---

[12] Again, I hasten to add that I am only referring to the general approach of deriving the distribution of events among microstates based on knowledge of average macroscopic properties. Classical statistical mechanics involves many substantive physical assumptions that do not apply to ecological communities.

species pool and *given these constraints and nothing else* (thus, maximizing relative entropy subject to these constraints) derive the resulting species abundance distribution.

If you go back to Chapter 4 you will recall that the basic method of maximum entropy is to (i) specify a maximally uninformative prior, which is determined by the nature of the possible states of the system, (ii) specify additional constraints on mean values of some properties of each element in the system, and (iii) maximize the relative entropy given (i) and (ii). Now, in Chapters 4 and 6 we were given a species pool and some properties (functional traits) of each species. In fact, it is because we did know the identity of each species that we could specify a species pool. Since we knew which species were in the species pool the goal was to predict the relative abundance of individuals in each species. The possible "states" in which an element could exist (i.e. an individual or unit of biomass) consisted of each possible species in the species pool. If you look at Figure 7.1 (top row) you will see that the abscissa, which lists the possible states (species), are fixed, discrete and unordered. We know (Chapter 4) that the maximally uninformative prior in such a case is a uniform distribution. Because the maximally uninformative prior is a uniform distribution, we can effectively ignore it and simply maximize the entropy (Chapter 4).

The problem is rather different when we are dealing with an unknown species pool. We can't possibly predict the proportion of individuals in each species since we don't know which species these are (or even how many they are). What we can possibly predict is the proportion of species having a given number of species; i.e. the species abundance distribution. However, if you look again at Figure 7.1 (bottom row) where the species abundance distribution is shown, you will notice that the states (i.e. the variable defining the abscissa) are not "species" but rather "numbers of individuals". A variable like the "number of individuals", although discrete,[13] is not fixed and also has a natural ordering. But we know that the maximally uninformative prior for such a variable is not a uniform distribution, but rather $q(n) = 1/n$, where n is the number of individuals. Pueyo *et al.* (2007) derive this prior using Jaynes' principle of transformational invariance and Dewar and Porté (2008) use a previous derivation from information theory to get essentially the same result.[14]

---

[13] If we measured abundance by biomass rather than by numbers of individuals, then the states would not be discrete either.

[14] They actually obtain $s(n) = 1/(1 + n)$ because they include the state of $s(n = 0)$; i.e. a species with no individuals.

So, what sort of species abundance distribution $S(n)$ would result if we assumed that there were absolutely no constraints on community assembly except for the logical requirement that there must be some finite number (N) of individuals in the community? If we let $s(n)$ be the probability of a species having n individuals, this results in the following constraint:

$$\sum_{n=1}^{n=\infty} s(n) = 1$$

$$N \sum_{n=1}^{n=\infty} s(n) = N(1). \qquad \text{(Eqn. 7.7a, b)}$$

Note that Equation 7.7b will hold for any N, so long as this value is finite. We are not (yet) assuming that the total number of individuals is fixed but simply that it is not infinite. We now maximize the relative entropy $RH = -\sum p_i \ln(p_i/q_i)$, subject to this single constraint (Chapter 4). The result is simply the geometric series (Equation 7.8), where λ_0 is the constant for normalization of the probability distribution. So, if there were no constraints during community assembly other than the requirement that the number of individuals in the community is finite then we would expect a geometric series:

$$S(n) = \frac{q(n)e^{-\lambda_0}}{\sum_i e^{\lambda_0}} = q(n)e^{-\lambda_0} - \frac{a_1}{n}. \qquad \text{(Eqn. 7.8)}$$

However, we can do better than this. In the neutral model we saw that, once a dynamic equilibrium is reached, the total number of individuals (N) and the total number of species (S) will tend towards constant values $(\hat{N}, \hat{S})$ that are determined by the rates of birth, death and immigration; the circumflex accent means that these are fixed by a dynamic equilibrium. By definition, the average number of individuals per species ($\bar{n}$) in any community is $\bar{n} = N/S$. Therefore, once a dynamic equilibrium has been reached for the total numbers of individuals $(\hat{N})$ and species $(\hat{S})$ in the community then this will also constrain the average number of individuals per species; i.e. $\bar{n} = \hat{N}/\hat{S}$. This is true whether the population dynamics are neutral or not, so long as a dynamic equilibrium has been reached. However, we can also write the average number of individuals per species as a constraint (Equation 7.9):

$$\sum_{n=1}^{\bar{N}} ns(n) = \bar{n} = \frac{\hat{N}}{\hat{S}}. \qquad \text{(Eqn. 7.9)}$$

If we maximize the relative entropy subject to both constraints (Equations 7.8 and 7.9) then we obtain Equation 7.10 which, incredibly, is Fisher's logseries distribution! If you go back to the discussion concerning the derivation of the neutral model for a metacommunity you will recall that the solution was indeed the logseries distribution and that this occurred once a dynamic equilibrium had been reached. We now have the same result without assuming neutrality but only a dynamic equilibrium such that the average number of individuals per species becomes constant:

$$S(n) = \frac{q(n)e^{-\lambda_0-\lambda_1 n}}{\sum_n e^{-\lambda_0-\lambda_1 n_0}} = q(n)\frac{e^{-\lambda_0}e^{-\lambda_1 n}}{Z} = \frac{a_2 e^{-\lambda_1 n}}{n}. \qquad \text{(Eqn. 7.10)}$$

So far, simply by assuming that our community has (i) a finite number of individuals and (ii) that it has reached a steady-state dynamic equilibrium with respect to the total number of individuals and the total number of species, we have been able to derive two well-known species abundance distributions. How about going a bit further?

As in Chapter 5, I begin with the definition of the per capita growth rate of any species i. This time, I deal with the entire regional species pool (with S_{max} species) and introduce an indicator function, $I(i, t)$, that takes a value of 1 if species i is present in the local community at time t, and that takes a value of 0 otherwise. The per capita growth rate of the average species in the local community, having $S(t)$ species at time t (thus, $S(t) = \sum_{i=1}^{S_{max}} I(i,t)$), is therefore

$$\bar{r}(t) = \frac{1}{S(t)}\sum_{i=1}^{S_{max}} S(t)r_i(t) = \frac{1}{S(t)}\sum_{i=1}^{S_{max}} I(i,t)r_i(t) = \frac{1}{S(t)}\sum_{i=1}^{S_{max}} I(i,t)\frac{1}{n_i(t)}\frac{dn_i(t)}{dt} = \frac{1}{S(t)}\sum_{i=1}^{S_{max}} I(i,t)\frac{d\ln(n_i(t))}{dt}.$$

These are simply mathematical identities and are true by definition. The discrete-time version of the per capita growth rate over one time interval is, by definition, $\ln(n_i(t+1)) - \ln(n_i(t))$ and so[15] $\bar{r}(t, t+1) = \frac{1}{S(t, t+1)}\sum_{i=1}^{S_{max}} I(t, t+1)(\ln(n_i(t+1)) - \ln(n_i(t)))$. Now consider the dynamics of this process over time, starting at time zero. Over the first (tiny) time interval we have $\bar{r}(0, 1) = \frac{1}{S(0,1)}\sum_{i=1}^{S_{max}} I(0,1)(\ln(n_i(1)) - \ln(n_i(0))) = \overline{\ln(n(1))} - \overline{\ln(n(0))}$ but this can be re-written as $\overline{\ln(n(1))} = \bar{r}(0, 1) + \overline{\ln(n(0))}$. Over the second (tiny) time interval we have $\bar{r}(1, 2) = \frac{1}{S(1, 2)}\sum_{i=1}^{S_{max}} I(1, 2)(\ln(n_i(2)) - \ln(n_i(1))) = \overline{\ln(n(2))} - \overline{\ln(n(1))}$ but

---

[15] The discrete and continuous time versions are related by

$$\frac{d \ln(n_i(t))}{dt} = \frac{1}{n_i(t)}\frac{dn_i(t)}{dt} = \lim(\Delta t \to 0)\frac{1}{n_i(t)}\frac{\Delta n_i(t)}{\Delta t} = \lim(\Delta t \to 0)\frac{\ln n_i(t+\Delta t) - \ln n_i(t)}{\Delta t}.$$

this can be written as $\overline{\ln(n(2))} = \bar{r}(1,2) + \overline{\ln(n(1))} = \bar{r}(1,2) + \bar{r}(0,1) + \overline{\ln(n(0))}$ After iterating over T time steps we get Equation 7.11:

$$\overline{\ln(n(T-1,T))} = \sum_{t=0}^{T} \bar{r}(t-1,t) + \overline{\ln(n(0,1))}. \qquad \text{(Eqn. 7.11)}$$

We are almost there, but we need one final step. This step is not conceptually difficult but requires a bit of mathematical manipulation. To start, let's write "$N(t)$" to mean the total number of individuals of all species that are present in the local community at time t and let's write "$R(t)$" to mean the per capita growth rate of the entire local community; I'm using the notation "R" rather than "r" simply to emphasize that we are talking about all individuals in the local community rather than those belonging to any one species. Then, since $N(t) = \sum_{i=1}^{S_{max}} I(t)n_i(t)$ and since $\Delta N(t) = \sum_{i=1}^{S_{max}} I(t)\Delta n(t)_i$ it follows that

$$R(t) = \lim(\Delta t \to 0)\frac{1}{N(t)}\frac{\Delta N(t)}{\Delta t} = \lim(\Delta t \to 0)\frac{1}{N(t)}\sum_{i=1}^{S_{max}} I(t + \Delta t)\frac{\Delta n(t)_i}{\Delta t}$$

$$R(t) = \lim(\Delta t \to 0)\frac{1}{N(t)}\sum_{i=1}^{S_{max}} I(t + \Delta t)n(t)_i\left(\frac{1}{n(t)_i}\frac{\Delta n(t)_i}{\Delta t}\right)$$

$$R(t) = \lim(\Delta t \to 0)\frac{1}{N(t)}\sum_{i=1}^{S_{max}} I(t + \Delta t)n(t)_i\left(\frac{\ln n_i(t + \Delta t) - \ln n_i(t)}{\Delta t}\right)$$

$$R(t) = \lim(\Delta t \to 0)\sum_{i=1}^{S_{max}} I(t + \Delta t)\frac{n(t)_i}{N(t)}\left(\frac{\ln n_i(t + \Delta t) - \ln n_i(t)}{\Delta t}\right)$$

$$R(t) = \overline{r(t)}.$$

This unattractive bit of mathematics assures us that the per capita growth rate of the entire local community at a given time, $R(t)$, is equal to the per capita growth rate of the average species in this local community, $\overline{r(t)}$ at this time. Using this fact, we can re-write Equation 7.11 to get Equation 7.12:

$$\overline{\ln(n(T-1,T))} = \sum_{t=0}^{T} R(t-1,t) + \overline{\ln(n(0.1))}$$

$$\left(\prod_{i=1}^{S} n_i(T-1,T)\right)^{\frac{1}{S(T-1,T)}} = e^{\sum_{t=0}^{T} R(t-1,t) + \overline{\ln(n(0.1))}} \qquad \text{(Eqn. 7.12a, b)}$$

Equation 7.12a says that the average of the logarithm of the number of individuals in the local community ($\ln(n)$) over the time interval from $T-1$ to T equals the sum of the community per capita growth rates up to time T plus the average $\ln(n)$ at the start. But the average of $\ln(n)$ is simply the geometric mean of n. In other words, the geometric mean abundance of the species present in the community during each time interval is *constrained*. Furthermore, once the total population size of the community reaches a dynamic equilibrium at time T then the community per capita growth rate (R) will be zero and the sum $(\sum_t R(t-1, t))$ will be constant. If, as Hubbell's "zero-sum multinomial" model of neutral communities assumes, the total number of individuals in the community is always constant, then $\sum_t R(t-1, t) = 0$ and the geometric mean abundance is constrained to equal the value at the start of the dynamic process. Adding this third constraint equation (Equation 7.12a) to the first two (Equations 7.7 and 7.9) and maximizing the relative entropy, we get a new equation for the species abundance distribution (Equation 7.13).

Before we look at Equation 7.13, I have to point out an unpleasant fact. We ecologists are much too parochial. Ever since Motomura (1932), Corbet (1941) and Fisher *et al.* (1943) first described species abundance distributions we ecologists have been looking for *ecological* explanations for their origin. This is reflected in the recent debate about whether species abundance distributions are due to neutral or niche processes. In arguing about the relative importance of these two alternate ecological processes in generating species abundance distributions we are implicitly assuming that, whatever the reason, it is an *ecological* one. However, as Preston (1950) first[16] pointed out to us, and has been recently repeated by Nekola and Brown (2007), *"species"* abundance distributions are simply an ecological example of a much more common pattern of abundance distributions. Nekola and Brown (2007) show, for example, that the neutral model of Equation 7.6 provides an excellent fit to the "abundance" distribution of North American precipitation classes, of stock volumes for all publicly traded US corporations, for song performances of *Cowboy Junkies* set lists, and citation frequencies for scientific papers. We ecologists have mistakenly placed the emphasis on the first, rather than the last, word of the phrase "species abundance distributions".

---

[16] Preston later documented rarity enriched neutral-like species abundance distributions for the service life of restaurant drink tumblers, the static fatigue of glass and other materials, and ages of first marriage for women in Denmark, the UK and the USA (Preston 1981).

This navel gazing by ecologists is matched by researchers in other fields. It is worth quoting from a paper by a computer scientist, in which the reasons for lognormal distributions in that field are debated (Mitzenmacher 2004): *". . . [T]he argument over whether a lognormal or power law distribution is a better fit for some empirically observed distribution has been repeated across many fields over many years. For example, the question of whether income distribution follows a lognormal or power law distribution also dates back to at least the 1950s. The issue arises for other financial models, as detailed in (Mandelbrot 1997). Similar issues continue to arise in biology (Jain and Ramakumar 1999), chemistry (Nakajima and Higurashi 1998), ecology (Sole et al. 1987, Allen et al. 2001), astronomy (Wheatland and Sturrock 1996), and information theory (Li 1996, Perline 1996). These cases serve as a reminder that the problems we face as computer scientists are not necessarily new, and we should look to other sciences both for tools and understanding."* Since such similar mathematical patterns occur in such disparate phenomena, the explanation for such patterns cannot be unique to the biological process of community assembly (Nekola and Brown 2007). In this context, it is interesting that Preston himself (1950) suggested that the convergence of the Boltzmann distribution of gases, the Pareto distribution of wealth distribution among people, and the species abundance distribution in ecological communities might be the result of "statistical mechanics" although he could not do more that point out the analogy.

I am just as parochial as the next ecologist. I was therefore surprised, but shouldn't have been, when I came across an unpublished manuscript[17] by Andrey Rostovtsev who is a physicist at the Institute for Theoretical and Experimental Physics in Moscow. In this manuscript he looked for a class of dynamic systems for which there is a constraint on the geometric mean value of the observables and thus leading to a power law distribution. He found that this constraint exists in any system in which the probability density function evolves over time as a function of the relative, rather than the absolute, change of the observable per time step. In other words, a constraint on the geometric mean exists when the rate of change of something (x) is proportional to the amount of x existing at that time. In other words if $\frac{dx(t)}{dt} \propto x(t)$ or, equivalently, if $\frac{dx(t)}{x(t)dt} = K$, where K is an arbitrary constant; but this is simply the definition of per capita growth rate of a population! If the dynamics of a system $x = \{x_1, x_2, \ldots x_S\}$ over time are governed in this way then, after iterating over many time steps, the constraint

---

[17] http://arxiv.org/abs/cond-mat/0507414

in Equation 7.12a holds. If the system consists of "species" whose growth is determined by per capita growth rates, and if the observable is the number of individuals[18] in each species, then the assembly of ecological communities should always be constrained by the geometric mean abundance[19] of its members, according to Equation 7.12. If the system consists of "investors" whose wealth is determined by rates of return or interest rates (the amount of new money that they make or lose is proportional to the amount of money that they have invested), and the observable is the amount of money they own, then the assembly of wealth should always be constrained by the geometric mean abundance as well. In fact, *every* dynamic system for which the rate of change of something (x) is proportional to the amount of x existing at that time (i.e. a feedback relationship) should be subject to this type of constraint. This is presumably why, not only ecological communities, but also so many other phenomena subject to feedback show characteristic abundance distributions.

Okay, let's get back to species abundance distributions. If we include our constraint on the geometric mean abundance along with our two previous constraints (constraints 7.8, 7.9 and 7.12) and again maximize the relative entropy subject to these three constraints, we get a species abundance distribution that Pueyo *et al.* (2007) call the generalized logseries (Equation 7.13):[20]

$$S(n) = \frac{Sq(n)e^{\lambda_1 n - \lambda_2 \ln(n)}}{\sum\limits_n e^{\lambda_1 n - \lambda_2 \ln(n)}} = q(n)Se^{-\lambda_0 - \lambda_1 n - \lambda_2 \ln(n)} = \frac{Se^{-\lambda_0 - \lambda_1 n - \lambda_2 \ln(n)}}{n}$$

$$S(n) = \frac{Se^{-\lambda_0 - \lambda_1 n}}{n^{1+\lambda_2}} = \frac{S\left(e^{-\lambda_0}\right)e^{-\lambda_1 n}}{n^{1+\lambda_2}}. \qquad \text{(Eqn. 7.13)}$$

It is important to stress that this equation is a logical consequence of any dynamic process for which (i) the number of observables is finite, (ii) the mean value of the observable variable is fixed and (iii) the change in the observable variable over time is a feedback relationship such that $\frac{1}{x}\frac{dx}{dt} = f(\bullet)$. All of these conditions hold during community assembly so long as the total number of individuals and the total number of species each tends to some equilibrium value. Neutral models have this property but so do non-neutral processes of community assembly.

---

[18] Or some other measure of abundance, such as biomass.
[19] "The appropriate average fitness of a lineage in the long term is therefore the geometric mean of its fitness in each generation" (Bell 1996). I wonder if there is any connection?
[20] Pueyo *et al.* (2007) do not provide a physical reason for constraining $\ln(n)$, instead deducing it from empirical fits to a Taylor expansion of a logseries distribution.

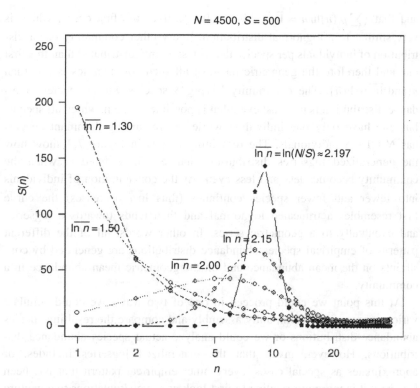

Figure 7.3. The behavior of the generalized logseries distribution for a fixed number of species (225) and a fixed geometric mean abundance (3.14) as the total number of individuals increases.

Let's explore this equation in more detail. First, notice that Equation 7.13 becomes the geometric distribution when $\lambda_1 \rightarrow 0$ and $\lambda_2 \rightarrow 0$ and that it becomes the logseries when $\lambda_2 \rightarrow 0$. However, for other combinations of these two parameters the equation approaches a lognormal distribution and even the neutral equation (Equation 7.6). This can be seen in Figure 7.3 where I model a community with $N = 4500$ individuals and $S = 450$ species and so $\bar{n} = 9$. The different species abundance curves were generated by keeping constant the total numbers of species and of individuals but changing the degree of evenness of the abundances of each species. Note that if a community is maximally even then all species will have the same abundance $n_i = (N/S)$ and so $\overline{\ln n} = \frac{1}{S}\sum_{i=1}^{S}\ln n_i = \frac{1}{S}\sum_{i=1}^{S}\ln \frac{N}{S} = \ln \frac{N}{S} = \ln \bar{n}$. Maximizing the entropy subject to the constraints that (i) the relative abundances sum to unity ($\sum p_i(n) = 1$), that the average number of individuals per species is 9 ($\sum p_i(n)n_i = \bar{n} = 9$), and that $\left(\sum p_i(n)\ln n = \overline{\ln n} = \ln \bar{n} = \ln 9\right)$, gives the first curve, which is very similar to a lognormal distribution. Every other community has a distribution of individuals per species that is less evenly distributed than this first one and therefore the geometric mean of all such communities is less than $\bar{n}$ (and $\overline{\ln n} < \ln \bar{n}$). The community having S species whose species abundance distribution is the least even that is possible is one in which all species but one have only one individual while the remaining dominant species has $N+1-S$ individuals. The remaining curves in Figure 7.3 show how the generalized logseries distribution changes, for a fixed $\bar{n}$, when the community becomes less and less even. As the concentration of individuals into fewer and fewer species continues (thus $\overline{\ln n}$ decreases) the curve first resembles a truncated lognormal and then tends towards a logseries and eventually to a geometric series. In other words, all of the different patterns of empirical species abundance distributions are generated by constraints on the mean abundance and on the geometric mean abundance in a community.

At this point we could proceed in one of two ways: we could simulate various processes of community assembly and compare the resulting species abundance distributions or we could analyze actual species abundance distributions. However, given that the generalized logseries includes, or approximates as special cases, every other empirical pattern that has been reported, it is more informative to first look at some simulations that capture some of these essential processes of community assembly and that span cases from neutral to non-neutral types.

Simulating demographic community assembly

All empirical studies of the species abundance distribution involve sampling from a spatial scale that is large enough to include many individuals. For instance, the data sets from Barro Colorado Island, Panama, that have been intensively studied have a scale of 40 square hectares. This means that there are likely different environmental conditions within the site, and this should be reflected in the simulation. Of course, in a strictly neutral model this is irrelevant since the fitness of individuals is equal and independent of environmental conditions. In non-neutral models this is important because it reflects the potential for niche partitioning. In order to be useful, a simulation should be simple enough so that one can understand what is happening while still containing the essential differences whose behavior is being studied.

I therefore chose a model very similar to the one used by Hubbell (2006). The purpose of this simulation model is to compare certain community-level patterns along a continuum of cases from strict neutrality (where the per capita probabilities of survival, reproduction, dispersal and immigration are the same for all species) to a strictly deterministic system in which survival, reproduction and dispersal are completely determined by the traits possessed by each species. Intermediate cases represent situations where the possession of traits simply shifts the probabilities towards or away from particular species.

In this simulation model the "site" consists of a series of 5000 cells arranged contiguously along a circle.[21] Each cell can contain at most one individual at a time and so the total number of individuals cannot exceed 5000. Each cell has two environmental properties: an environmental attribute (the "resource supply rate") that can take values from 0 to 40, and a probability of experiencing a disturbance event (i.e. density-independent mortality). A non-zero probability of disturbance means that a given cell can potentially be empty and so the actual total number of individuals in the community can be less than 5000; the actual number will be determined by the probabilities of disturbance, of immigration, of seed production and by the effects of sampling variation. The allocation of the environmental value to the cells can be (i) uniform and independent for each cell, resulting in approximately equal numbers of fine-grained microsite environments, (ii) uniform and spatially correlated, resulting in a continuously changing environmental gradient with a strong spatial signal or (iii) a distribution from a beta distribution such that, for various combinations of its two parameters, one can obtain a site with most cells being similar but with a small number of divergent environments; these environmental values can be spatially independent or correlated. A disturbance occurring in a cell means that the occupying individual (if it exists) is killed and this leaves the cell empty until the next dispersal or immigration event.

Each species in this simulation has three attributes: (i) a trait that determines its "resource use", (ii) the number of seeds produced per individual, and (iii) a dispersal trait that determines the chances of such seeds moving different distances (i.e. cells) left or right of its parent.

So, in this scenario the environment is characterized by two properties (the rates of resource supply and of density-independent mortality) and each individual is characterized by three traits (the ability to take up resources, the number of seeds produced and the average distance (measured in cells from the parent) that a seed can move). The ecological processes that determine which species will occupy which cell are competition and dispersal.

---

[21] A circle ensures that there are no boundaries while still having a finite number of cells.

Competition. When an individual "disperses" to a cell that is already occupied then either it (the invader) or the individual currently occupying the cell (the resident) will win the contest. If the resident wins then it remains in the cell. If the invader wins then the resident "dies" and the invader takes its place. If the cell is currently empty then the invader always wins. In the strictly neutral case both resident and invader have equal chances of winning and so the probability of the invader winning is 0.5. In order to represent different degrees of non-neutrality I introduce a weight that shifts this probability away from 0.5 by differing degrees depending on the trait value possessed by each individual relative to the environmental value in the cell.

This weight is determined by the degree of adaptation of each contestant in the cell. The degree of adaptation (a_{ij}) of an individual of species j to the environment at cell i (e_i) (the resource supply rate) is determined by the value of its "resource use" trait (t_j). Its level of adaptation is measured as the difference between the resource supply rate of the cell and the resource use trait of the species: $a_{ij} = \sqrt{(t_j - e_i)^2}$. A value of $a_{ij} = 0$ means that individual j is perfectly adapted to the environment in cell i (its resource use perfectly matches the resource supply). Since both the environment (e_i) and the trait take values between 0 and 40 an individual with the poorest fit to its cell will have $a_{ij} = 40$ while a perfectly adapted individual will have $a_{ij} = 0$. The difference in competitive ability between the two individuals (j and k) that are contesting cell i is measured as $d_{jk|i} = |a_{ij} - a_{ik}|$. In other words, the better-adapted individual is the better competitor and the greater the difference in adaptation the greater the difference in competitive ability. The probability, $p(j,k|i)$, of individual j winning the contest against individual k in the environment of cell i, where individual j is the better-adapted individual, is given by: $p(j, k|i) = 0.5 + 0.5\left(1 - e^{1 w d_{jk|i}}\right)$ and where $p(k, j|i) = 1 - p(j, k|i)$. Here "$w$" is the weight that determines the degree of deviation from neutrality. A strictly neutral model means that $w = 0$ and that $p(j, k|i) = p(k, j|i) = 0.5$. In a strictly neutral model the individuals still have trait differences but such differences in traits have no effect on survival. In scenarios that deviate from strict neutrality to various degrees the w values are greater than zero. Non-zero values of w mean that the probability that the individual from species j (i.e. the better-adapted species) will win increases above 0.5 and that the amount by which species j is favored will depend on the difference in the adaptive value between the two individuals and on the weight. Figure 7.4 shows how the probability of winning shifts in favor of species j for different values of w and different degrees of difference in the adaptive values of the two individuals.

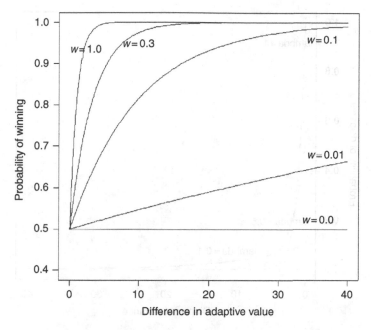

Figure 7.4. The function defining the probability of the better-adapted individual out-competing the more poorly adapted individual as a function of the difference in the adaptive values of the two and on the weight (w). A value of $w = 0$ corresponds to strict neutrality, a small value ($w = 0.01$) gives only a small advantage if the two individuals are close in adaptive value, while a large value ($w = 1.0$) results in even a slightly better-adapted individual almost always winning.

Dispersal. The probability that an individual that is currently residing in cell i will disperse a propagule to cell j around the circle (both forwards or backwards with equal probability) is defined by an exponential distribution. An exponential distribution has a single parameter (λ) and this parameter (the dispersal trait) is determined for each individual based on its species-specific dispersal ability and a weighting value according to: $\lambda_k = \lambda + w(l_k - \lambda)$, where l_k is a species-specific lambda value. If $w = 0$ then all individuals of all species have the same lambda value (λ) and so each individual can disperse to varying distances with exactly the same probability. Small values of w result in only slightly different dispersal abilities between individuals while $w = 1$ results in the dispersal ability of an individual being entirely determined by its species-specific lambda (l_j): $p(d_{ij}) = \lambda_k e^{-\lambda_k d_{ij}} = (\lambda + w(l_k - \lambda))e^{-(\lambda + w(l_k - \lambda))d_{ij}}$.

Note that if $w = 0$ then dispersal ability is strictly neutral ($p(d_{ij}) = \lambda e^{-\lambda d_{ij}}$) and if $w = 1$ then dispersal is strictly deterministic and defined by l_k, the species-specific dispersal trait ($p(d_{ij}) = l_k e^{-l_k d_{ij}}$).

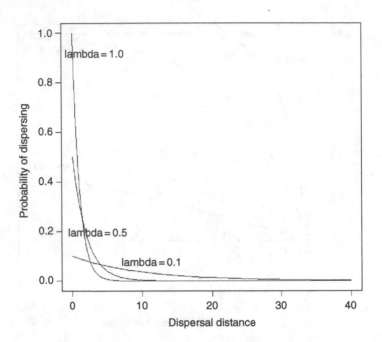

Figure 7.5. The probability of a seed dispersing a given distance from the parent for different values of lambda.

In the simulation the values of l_k are drawn from a lognormal distribution with mean of log(0.4) and standard deviation of 0.6; thus 95% of species will have an l_k value of between 0.12 and 1.30. Figure 7.5 shows how the probability of dispersing varies with the lambda parameter.

In the simulations to follow, the lambda value is set at 0.4 and so there are no systematic differences between species for this trait. This insures that almost all interactions are local since 95% of all dispersal events will be within 8 cells from the parent; the probability of dispersing within a distance d_{ij} is given by $1 - e^{-\lambda d_{ij}}$.

Seed production. Since seed production is usually normally distributed on a log scale, the number of propagules that an individual will disperse at any one time is given by a lognormal function:

$$p\left(n;\mu,\sigma^2\right) = \frac{1}{\sqrt{2\pi\sigma^2}} e^{-\frac{(\ln(n)-\mu)^2}{2\sigma^2}}.$$

In the neutral case the parameter values are $\mu = \log(5)$ and $\sigma = 1$. In order to simulate different degrees of nonnormality with respect to the average number of seeds per plant produced by different species, I again introduce a

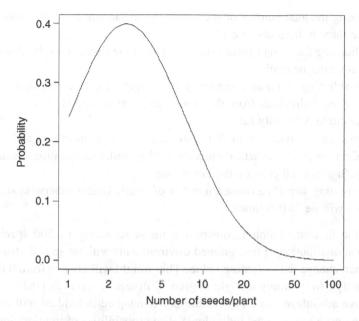

Figure 7.6. The probability of producing different numbers of seeds per plant.

weighting value and define $\mu_k = \mu + w(m_k-\mu)$, where m_k is the trait value determining the mean value for species k. In the simulations to follow the value of m_k is chosen from a lognormal distribution whose mean is $\log(5)$ and whose standard deviation is 1; thus 95% of species will have values between 0.7 and 35.5 seeds/individual. Neutrality occurs when $w = 0$ while $w = 1$ corresponds to the case where the trait completely defines the seed production. Again, there are no systematic differences between species in μ and σ in the simulations to follow. Figure 7.6 shows the probability function for seed production in the neutral case. As can be seen, the average number of seeds per plant is ~2.7 and ±1SD corresponds to between 1 and ~7 seeds per plant.

Metacommunity. The abundance distribution of individuals in the meta-community is determined by a geometric series. By choosing different values of the constant one can get different levels of evenness. In the following series of simulations I use a rather uneven distribution with a few common species and many rare species since this is the most common pattern in real communities.

A simulation run consists of the following.

(1) Allocating environmental values to each cell in order to establish the underlying environmental gradient.
(2) Allocating probabilities of disturbance to each cell.

(3) Fixing the total number of species in the metacommunity and allocating the three trait values to each.

(4) Choosing the weight values that determine the distance of the dynamics from strict neutrality.

(5) Establishing an initial community by having all cells empty and then allowing individuals from the metacommunity to migrate into cells until the circle is initially full.

(6) Iterating the dynamics of disturbance, dispersal, competition and migration from the metacommunity, cell by cell. One iteration consists of cycling over all cells on the circle once.

(7) Repeating step (6) a chosen number of times. Unless otherwise stated, this will be 25 000 times.

In the first simulation I construct a metacommunity of 500 species, a uniform distribution of fine-grained environments without spatial structure, and then change the weighting value. This weight will change from 0 (strict neutrality), to 0.3 (only a slight degree of fitness differences leading to an adaptive advantage), to 1 (where the better-adapted individual will always win against a less-adapted individual). The probabilities of immigration and disturbance are 0.05. Therefore, on average, 5% (250 cells) will experience a density-independent death at any time and the same number of cells will independently receive an immigrant from the metacommunity. Figure 7.7 shows the results.

There are a number of things to notice in Figure 7.7. First, notice how the effects of disturbance and immigration result in individuals with a wide range of trait values being found at the same environmental position. More importantly, this occurred even in the most extreme case of non-neutrality; that is, when competitive ability was completely determined by the "resource use" trait value of the individual that confers adaptation to the environment (Figure 7.7i). Of course, in a perfectly neutral community there is no relationship between traits and environments (Figure 7.7c). Second, notice that the generalized logseries distribution always gave a better fit to the data than did the logseries distribution. Third, notice how the effect of non-neutrality on the species abundance distribution is most obvious for the abundant species, not for the rare species.

Next, I simulated community assembly by changing three things. First, I increased the number of species in the metacommunity from 500 to 1000. Next, I reduced the disturbance rate from 0.05 to 0.01 (i.e. only 1 in 100 cells will, on average, experience density-independent mortality). Finally, I changed the distribution of environmental conditions within the

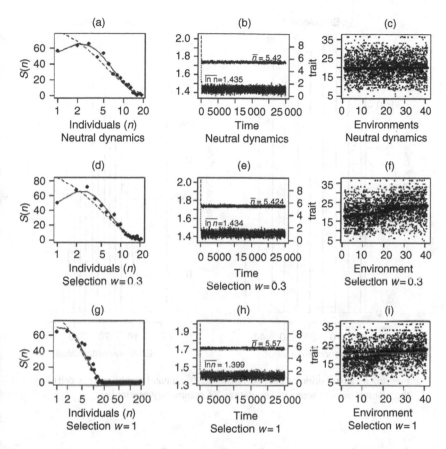

Figure 7.7. The results of simulations for a perfectly neutral (top row, (a)–(c)), a slightly non-neutral (middle row, (d)–(f)) and a strongly non-neutral (bottom row ((g)–(i)) community. Figures (a), (d) and (g) show the species abundance distribution (points) along with the generalized logseries (solid line) and the logseries distributions (broken line). Figures (b), (e), and (h) show the dynamics of the values of $\bar{n}$ and $\overline{\ln n}$ over time. Figures (c), (f) and (i) show the trait value related to adaptive competitive ability of each individual relative to the environmental value of the cell in which the individual resides; the solid line is the community-aggregated value at each environmental value.

site. In the first simulations the site was a fine-grained one with approximately equal numbers of microsites (i.e. cells) having each level of "nutrient supply" (i.e. the environmental variable) and randomly distributed without any spatial structure. In this second series of simulations the site is still fine-grained and without any spatial structure but some levels of the environmental variable are much more common than others. The

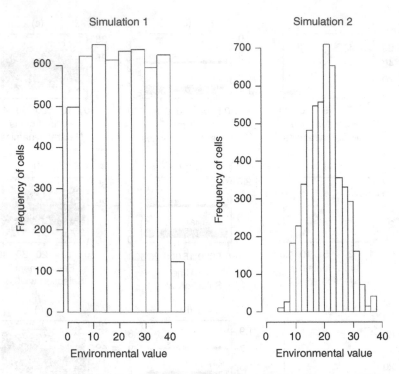

Figure 7.8. Distribution of microsites (i.e. environmental conditions per cell) in the first and second series of simulations of community assembly.

distribution of microsites is shown in Figure 7.8 and the results of the simulation are shown in Figure 7.9.

The effects of non-neutrality are now more pronounced since the correlation between traits and environments (Figure 7.9d, i) is stronger. This is because the disturbance rate is lower, thus increasing the importance of competition relative to immigration or seed dispersal in determining occupancy of a cell. With twice as many species in the species pool (1000 vs. 500) the distribution of abundance is less even, as shown by the mean of $\ln(n)$. The resulting species abundance distributions are still well described by the generalized logseries.

The last series of simulations are like the second one except that there is now a strong spatial structure to the environmental values. Cells close together have similar environmental values and the spatial gradient changes smoothly around the circle, resulting in a sine wave when "rolled out". Figure 7.10 shows the distribution of the environmental variable over cells

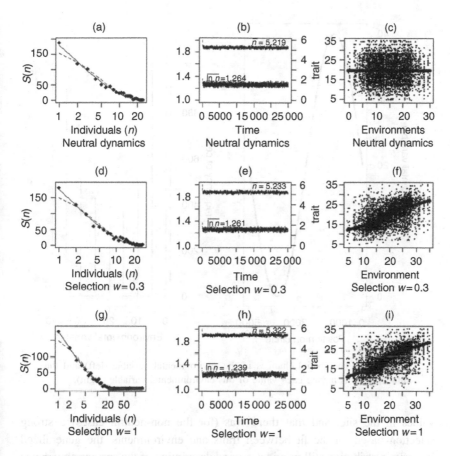

Figure 7.9. The results of simulations for a perfectly neutral (top row, (a)–(c)), a slightly non-neutral (middle row, (d)–(f)) and a strongly non-neutral (bottom row ((g)–(i)) community. Figures (a), (d) and (g) show the species abundance distribution (points) along with the generalized logseries (solid line) and the logseries distributions (broken line). Figures (b), (e), and (h) show the dynamics of the values of $\bar{n}$ and $\overline{\ln n}$ over time. Figures (c), (f) and (i) show the trait value related to adaptive competitive ability of each individual relative to the environmental value of the cell in which the individual resides; the solid line is the community-aggregated value at each environmental value. In this simulation there is 5 times less density-independent mortality, there is twice as many species in the meta-community, and the distribution of environmental conditions is very uneven.

and as a function of the distance around the circle and Figure 7.11 shows the result of the simulation.

Despite the fact that there was a strong spatial structure in the underlying environmental variable, that there was an uneven distribution of environmental

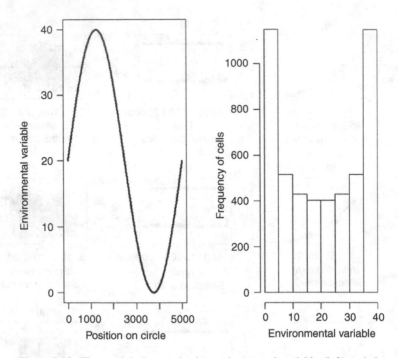

Figure 7.10. The spatial changes in the environmental variable (left) and the number of cells having each value of this environmental variable (right).

values in the site, and that there was (for the non-neutral cases) a strong selection based on the fit between traits and environments, the generalized logseries distribution still provides a good description of the species abundance distribution. In fact, when we look at some empirical examples, you will see that the pattern in this decidedly non-neutral case is very similar to such empirical patterns that have been successfully fit using the neutral equation (Equation 7.6). This does not mean that the generalized logseries distribution, derived from two general constraints on population dynamics, will always work. The generalized logseries distribution will work when these are the *only* systematic constraints.

Finally, how does the generalized logseries distribution perform on empirical species abundance distributions. Volkov *et al.* (2003, 2005) fit the neutral equation (Equation 7.6) to observed species abundance distributions from six different tropical forests, including the Barro Colorado Island site that was the inspiration for Hubbell's (2001) neutral model. In Figure 7.12, I show these six empirical species abundance distributions and superimpose on these empirical values the predicted values from (i) the generalized logseries

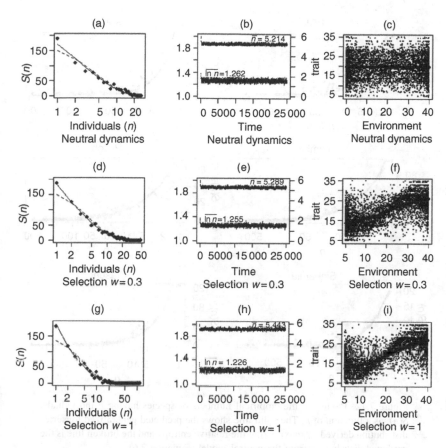

Figure 7.11. The results of simulations for a perfectly neutral (top row, (a)–(c)), a slightly non-neutral (middle row, (d)–(f)) and a strongly non-neutral (bottom row ((g)–(i)) community. Figures (a), (d) and (g) show the species abundance distribution (points) along with the generalized logseries (solid line) and the logseries distributions (broken line). Figures (b), (e), and (h) show the dynamics of the values of $\bar{n}$ and $\overline{\ln n}$ over time. Figures (c), (f) and (i) show the trait value related to adaptive competitive ability of each individual relative to the environmental value of the cell in which the individual resides; the solid line is the community-aggregated value at each environmental value. This simulation was like the previous one (Figure 7.9) except that there was a strong spatial structure to the arrangement of environmental values.

(solid line) based on the Maximum Entropy Formalism and (ii) the neutral model of Equation 7.6 (broken line). These curves are statistically indistinguishable. Note that the neutral equation would not fit the neutral pattern in Figure 7.7 very well at all because, in that graph, the curve dips at the lowest abundance level, resulting in a lognormal-like curve.

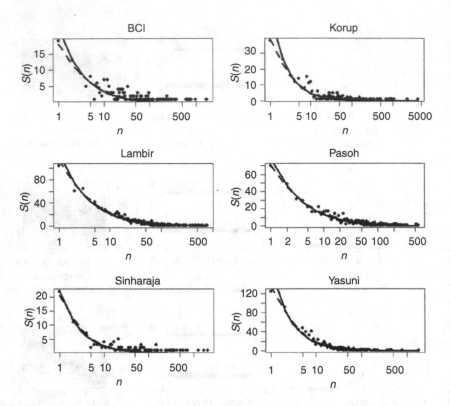

Figure 7.12. Points are the empirical numbers of species having n individuals, $S(n)$, as a function of n. The solid line shows the predicted generalized logseries distribution derived by maximizing the relative entropy and the broken line is the predicted distribution from the neutral model (Equation 7.6).

A resource-based statistical mechanics

When we considered a statistical mechanistic model based on population dynamics the result was a generalized logseries distribution. The only empirical information needed was $\bar{n}$ and $\overline{\ln n}$ (and the total number of species at the site if you want the number of species rather than the relative number of species). However, such a distribution (like the geometric, logseries and the neutral distributions) is not really very useful in any predictive sense since there is no obvious independent way of obtaining the empirical information besides actually counting (or sampling) the number of individuals per species. Dewar and Porté (2008) took a rather different, but still statistical mechanistic, approach. Rather than concentrating on population dynamics and the constraints that arise from them, they studied what would happen if there were only two "conservation" laws at work. The first conservation law was our familiar constraint that the total number of individuals in the community must be finite; thus $f_1(n) = \sum_{i=1}^{S} n_i = N$. The second conservation law was that the total amount of some limiting resource (R) must be fixed; if u_i is the amount of this resource taken up by an average individual of species i then $f_2(n) = \sum_{i=1}^{S} n_i u_i = R$. Given these two conservation laws one can specify two constraints[22] (Equation 7.14):

$$\sum_n p(n) \sum_{i=1}^{S} n_i = \sum_n p(n) N = N,$$

$$\sum_n p(n) \sum_{i=1}^{S} n_i u_i = \sum_n p(n) R = R. \qquad \text{(Eqn. 7.14)}$$

The solution that Dewar and Porté (2008) were looking for is slightly different than the one of Pueyo *et al.* (2007). Dewar and Porté considered S to be the number of species in the entire metacommunity and wanted the probability of observing a particular distribution $p(\mathbf{n})$ of observing n_1 individuals of species 1, n_2 individuals of species 2, and so on; so $p(\mathbf{n}) = p(n_1, n_2, \ldots, n_S)$. Note that a species can be present in the metacommunity but absent in the local community so that some of the n_i values in $p(\mathbf{n})$ can be zero. This is different from what we considered above based on Pueyo *et al.* (2007) where the identity of species was ignored and we were only looking for the probability of actually observing S species in the community having n individuals. Since it is impossible to observe species in the local community that have $n = 0$ individuals (i.e. species that are not there), we did not have to consider the case of $n = 0$. Since we now have to consider cases in which, for example, species 1 is missing from the local community, we need to consider cases in which $p(\mathbf{n}) = p(0, n_2, \ldots, n_S)$. We therefore need to consider a slightly different maximally uninformative prior.[23] In the absence of any constraints except for the definition of an "individual" (i.e. an integer number), the number (n) of individuals of species i can take any of the infinite set of values $n_i = 0, 1, 2, \ldots \infty$. Letting $m_i = n_i + 1$, this means that m_i can take any value in the set $m_i = 1, 2, \ldots \infty$. Now we use our maximally uninformative prior $q(m) = \frac{1}{m} = \frac{1}{n+1}$.

---

[22] Of course, we must also include the logical constraint that the probabilities sum to unity: $\sum_n p(n) = 1$.

[23] Dewar and Porté actually obtain the maximally uninformative prior from an approximation of the result in Rissanen (1983).

Well, you know what comes next. Now that we have our constraints, we must maximize the relative entropy $(-\sum_n p(\mathbf{n})\ln\frac{p(\mathbf{n})}{q(\mathbf{n})})$ subject to these constraints, given the maximally uninformative prior $q(\mathbf{n})$. We do this (Chapter 4) by solving for the Lagrangian multipliers (λ) in the objective function $0 = \sum_n p(n)\ln\frac{p(\mathbf{n})}{q(\mathbf{n})} - \lambda_0(1 - \sum_n p(\mathbf{n})) - \lambda_1(N - \sum_{\mathbf{n}} p(\mathbf{n})\sum_{i=1}^{S} n_i) - \lambda_2(R - \sum_{\mathbf{n}} p(\mathbf{n})\sum_{i=1}^{S} n_i u_i)$. The notation in this objective function might be a bit confusing so please note that $\sum_n p(n) = \sum_{n_1,n_2,\cdots,n_s=0}^{\infty} p(n_1, n_2, \cdots n_s)$; that is, the summation is from 0 to infinity for each species in the metacommunity. This might seem like a miserable looking equation to solve, but its solution is simply the same generalized exponential distribution that you already know. Equation 7.15 is the result:

$$p(\mathbf{n}) = \left(\frac{1}{n+1}\right)(e^{-\lambda_0})\left(e^{-\lambda_1 \sum_{i=1}^{S} ni - \lambda_2 \sum_{i=1}^{S} n_i u_i}\right)$$

$$= \left(\frac{1}{Z}\right)\left(\frac{1}{n+1}\right)e^{-\sum_{i=1}^{S} n_i(\lambda_1+\lambda_2 u_i)}$$

$$p(n) = \frac{1}{Z(n+1)}\prod_{i=1}^{S} e^{-n_i(\lambda_1+\lambda_2 u_i)}. \qquad \text{(Eqn. 7.15)}$$

Evaluating this equation requires a bit more work, which we will do in a moment, but let's stop for a moment to compare Equation 7.15 with the type of equations for the species abundance distribution that we have seen before. In the generalized logseries we must know the values of the average number of individuals per species ($\bar{n}$, therefore the total number of species S and the total number of individuals N) and $\overline{\ln n}$. In any practical sense the only way of obtaining such information is to sample the community. But if we have to sample the community then we will already have an estimate of the species abundance distribution! In the model of Dewar and Porté (Equation 7.14) we only need to know the total number of individuals and the total amount of the resource in the community (N and R). In order to get values for Equation 7.15 we also need to know the values of the per capita amount of resource captured by species i (i.e. u_i) and this means (since we are imagining a metacommunity with an arbitrary number of species whose identity is unknown) that we must know the distribution of the per capita resource use in the metacommunity. We have come back (indirectly) to functional traits since per capita resource use will be determined by things like adult size or growth rates. More about this later, but let's first get back to evaluating Equation 7.15.

The joint probability, $p(\mathbf{n})$, of this maximum entropy solution is the product of the probability distributions for each species; if this wasn't true then there would have to be additional constraints of the covariances of the species' abundances which we haven't included. The probability that species i will have n individuals is therefore[24] $p_i(n) = \frac{1}{Z_i(n+1)} e^{-n(\lambda_1+\lambda_2 u_i)}$. Z_i is $\sum_{n=0}^{\infty} \frac{e^{-n_i(\lambda_1+\lambda_2 u_i)}}{n+1} = \sum_{n=0}^{\infty} \frac{x_i^n}{n+1}$ and Dewar and Porté show that this infinite series converges to $Z_i = \frac{|\ln(1-x_i)|}{x_i}$. From this they obtain the final result (Equation 7.16):

$$p_i(n) = \frac{1}{|\ln(1 - e^{-\lambda_1-\lambda_2 u_i})|(n+1)} e^{-(n+1)(\lambda_1+\lambda_2 u_i)}$$

$$= \frac{x_i^{n+1}}{|\ln(1-x_i)|(n+1)}. \qquad \text{(Eqn. 7.16)}$$

Equation 7.16 gives the probability of species i, with a per capita resource use of u_i, having n individuals. In order to get the species abundance distribution, we need the *average* number of species with abundance n and this is given by $\sum_{i=1}^{S} p_i(n)$. In order to get this value we must know the distribution of per capita resource use (u_i) in the species pool (the metacommunity).

Back to plant traits

In the equation of Dewar and Porté (2008) we see the term "u_i", which is the amount of resources captured by the average individual of species i. Notice that "*the amount of resources captured by the average individual of species i*" is a trait of species i. It is not strictly a morphological or physiological trait but is the consequence of such traits. We could write u_i as a function of such traits, $u_i = f(t_{i1}, t_{i2}, \ldots)$ but, to keep the algebra simple, I will leave the trait as u_i. Now imagine that one researcher has determined the entire species pool of the metacommunity, has measured the per capita resource use (u_i) of each of the (known) S species in the pool and (again, to keep the algebra simple) imagine that this single "trait", u_i, is sufficient to determine the fitness of the species and, thus, its relative abundance. This is the situation described in Chapter 4, and we know from Chapter 4 that the predicted relative abundance of each species in the species pool at the site

[24] $Z = \prod_{i=1}^{S} Z_i$ since $Z = \sum_{n_1,n_2,\ldots n_S} \prod_{i=1}^{S} \frac{e^{-n_i(\lambda_1+\lambda_2 u_i)}}{n+1} = \prod_{i=1}^{S} Z_i$.

will be given by $p_i = e^{-a-\beta u_i}$, where a and β are the Lagrangian multipliers. If there are a total of N individuals at the site, the predicted number of individuals (i.e. abundance) of species i will be $\hat{n}_i = Np_i = Ne^{-a-\beta u_i}$. Now imagine that a different researcher, who does not know which species are in the species pool, takes the approach of Dewar and Porté and uses Equation 7.16. The average (i.e. expected) number of individuals of an arbitrary species with a per capita resource use of u_i is, from Equation 7.16, $\bar{n}_i = \sum_{n=0}^{\infty} np_i(n) = \sum_{n=1}^{\infty} np_i(n)$ but this is $\bar{n}_i = \frac{1}{\left|\ln\left(1-e^{-\lambda_1-\lambda_2 u_i}\right)\right|}\sum_{n=1}^{\infty}\frac{n}{n+1}e^{-(n+1)(\lambda_1+\lambda_2 u_i)}$. Notice that both researchers are predicting the same thing and from this we could derive the relationship between the two sets of Lagrangian multipliers (a, β) and (λ_0, λ_1). For those who are interested, Box 7.1 gives the relationship between the model of Dewar and Porté (2008) and the model developed in this book.

Box 7.1. Relating models for fixed and unknown species pools

The Taylor series expansion of $\log(1+x)$ about the value $x = 0$ is given by $\ln(1+x) = x - \frac{x^2}{2} + \frac{x^3}{3} - \cdots$. The series expansion converges in the range $k - 1 \prec x \prec 1$ (successive terms get smaller). We have $\ln(1-x) = -x - \frac{x^2}{2} - \frac{x^3}{3} - \cdots$ and $-\ln(1-x) = x + \frac{x^2}{2} + \frac{x^3}{3} + \ldots$ and these expressions are also valid in the range $-1 \prec x \prec 1$. We can use this to write

$$Z_i = \sum_{n=0}^{\infty}\frac{x_i^n}{n+1} = \frac{1}{x_i}\sum_{n=0}^{\infty}\frac{x_i^{n+1}}{n+1} = \frac{1}{x_i}\left(x_i + \frac{x_i^2}{2} + \frac{x_i^3}{3} + \cdots\right) = \frac{-\ln(1-x_i)}{x_i}.$$

provided $-1 \prec x_i \prec 1$. Now the definition of x_i is $x_i = e^{-(\lambda_1+\lambda_2 r_i)}$ so automatically we have $x_i \succ 0$; the convergence criterion is therefore $x_i \prec 1$. Since $x_i \succ 0$ we have $\ln(1-x_i) \prec 0$ so that $-\ln(1-x_i) \succ 0$ and we can write $-\ln(1-x_i) = |\ln(1-x_i)|$ giving the final result

$$Z_i = \frac{|\ln(1-x_i)|}{x_i}.$$

Now, we need the expected value of $p_i(n)$; i.e. the expected number of individuals of a species i having u_i. This is

$$\sum_{n=0}^{\infty} np_i(n) = \frac{1}{\left|\ln\left(1-e^{-\lambda_1-\lambda_2 r_i}\right)\right|}\sum_{n=0}^{\infty}\frac{n}{n+1}e^{-(n+1)(\lambda_1+\lambda_2 r_i)}.$$

This can be calculated by re-writing the infinite series as the difference of two other series:

$$\sum_{n=0}^{\infty}\frac{n}{n+1}x_i^{n+1} = \sum_{n=0}^{\infty}\frac{(n+1)-1}{n+1}x_i^{n+1} = \sum_{n=0}^{\infty}x_i^{n+1} - \sum_{n=0}^{\infty}\frac{1}{n+1}x_i^{n+1}.$$

The first series on the right-hand side is

$$\sum_{n=0}^{\infty}x_i^{n+1} = x_i\sum_{n=0}^{\infty}x_i^n = x_i\left(1+x_i+x_i^2+x_i^3+\cdots\right) = \frac{x}{1-x_i},$$

which uses the Taylor series expansion of $(1-x)^{-1}$ valid for $-1 \prec x \prec 1$. The second series on the right-hand side is

$$\sum_{n=0}^{\infty}\frac{1}{n+1}x_i^{n+1} = x_i + \frac{x_i^2}{2} + \frac{x_i^3}{3} + \ldots = -\ln(1-x_i) = |\ln(1-x_i)|.$$

It then follows that the expected number of individuals of species i is

$$\bar{n}_i = \sum_{n=0}^{\infty}np_i(n) = \frac{1}{|\ln(1-x_i)|}\sum_{n=0}^{\infty}\frac{n}{n+1}x_i^{n+1} = \frac{x_i}{1-x_i}\frac{1}{|\ln(1-x_i)|} - 1.$$

In order to work out the relationship between the equation of Dewar and Porté (2008) (for a species pool having an unknown number of species) and the equation used in this book (for a fixed and known number of species in the species pool), we first note that "r_i" (the per capita resource use of species i) is a species trait. The probability of observing a species in the (fixed and known) species pool having trait r_i is $p_i = e^{-a-\beta r_i}$, and p_i can be interpreted as a predicted relative abundance. Thus the predicted abundance (i.e. number of individuals $\hat{n}_i$) in a community with a total of N individuals would be $Np_i = Ne^{-a-\beta r_i}$. The predicted abundance $\hat{n}_i$ should be close to the average abundance of Dewar and Porté (2008) in a large sample; i.e.

$$Ne^{-a-\beta r_i} \approx \sum_{n=0}^{\infty}np_i(n) = \frac{1}{|\ln(1-e^{-\lambda_1-\lambda_2 r_i})|}\sum_{n=0}^{\infty}\frac{n}{(n+1)}e^{-(n+1)(\lambda_1+\lambda_2 r_i)}.$$

The correspondence of this equation[25] with the model developed in this book emerges in the limit $x_i \to 0$, which corresponds to the situation where each $\bar{n}_i$ is small (i.e. large community resource-use $\bar{R}$, small total population $\bar{N}$). The expression for $\bar{n}_i$ in the model developed in this book then reduces to the simpler result $\bar{n}_i \approx \frac{x_i}{2} = \frac{1}{2}e^{-\beta(r_i-\mu)}$ when $x_i \ll 1$. This

---

[25] This section is developed from personal correspondence with R. Dewar.

Box 7.1. (Continued)

can be derived by inserting the series expansion of $-\log(1-x_i)$ and studying the approximate behavior of $\bar{n}_i$ for small values of x_i. Therefore $\bar{N} = \sum_i \bar{n}_i \approx \sum_i \frac{x_i}{2} = \frac{1}{2}e^a e^{\beta\mu}$ where I have defined $e^a \equiv \sum_i e^{-\beta r_i}$, and so we can write the approximation for $\bar{n}_i$ as $\bar{n}_i \approx \bar{N}e^{-a-\beta r_i}$.

The correspondence that emerges here is entirely analogous to the way Bose–Einstein statistics reduce to Maxwell–Boltzmann statistics in the limit of high temperature or low particle concentration.[26]

---

[26] See http://en.wikipedia.org/wiki/Maxwell–Boltzmann_statistics.

8

Epilogue: traits are not enough

Throughout this book I have contrasted "neutral" processes with "selective" processes that are based on trait-based environmental filtering. If "neutral" is interpreted in its strict sense of requiring exactly equivalent fitness between all individuals of all species then community assembly is either neutral or it is not. As I have already argued, such a strict conception of neutral community assembly makes no sense, except as a benchmark to measure departures, because it makes natural selection impossible. If "neutral" is interpreted more loosely, as any process that is independent of functional trait differences and that affects population dynamics by affecting realized rates of survival, reproduction or immigration, then community assembly almost surely occurs through an interplay of both "neutral" and "selective" processes. This looser meaning of "neutral" would include the stochastic demographic effects of birth, death and immigration that are especially important when population sizes are small but would also include all those dispersal limitations, including landscape features and history, which prevent propagules from moving between local communities and which are not related to functional traits. In what follows I will use "neutral" in this less rigid sense. In real ecological communities traits are not enough because such neutral processes always exist. Some communities might be predominantly structured by neutral processes while others might be predominantly structured by selective processes. The relative importance of these two groups of processes will determine whether or not the model developed in this book will have good predictive ability or not. I therefore want to finish this book by emphasizing this interplay and suggesting some ways that the model might be modified. To do this, I need to explain why we might actually want to use priors having somewhat more information than the maximally uninformative one discussed so far. For lack of a better term, I will call this a "neutral" prior.

Neutral, rather than maximally uninformative, priors

Let's remind ourselves of what we are trying to do. We want to translate the verbal model of trait-based environmental filtering into mathematical form. The mathematical model that we developed in Chapter 4, based on the Maximum Entropy Formalism, does this *in principle* by linking macroscopic constraints to Bayesian probabilities. Such a mathematical model reflects trait-based environmental filtering when the macroscopic constraints are expressed in the form of the community-aggregated traits because the filtering is due to natural selection and natural selection (via the Breeder's Equation) places constraints on community-aggregated traits. In *practice*, however, we might be setting the bar too high when starting with a maximally uninformative prior. Why?

If you look again at the definition of relative entropy (Equation 4.1) you will see that maximizing it subject to the macroscopic constraints is equivalent to choosing a posterior probability distribution (**p**) that is as close as possible to the prior distribution (**q**) while excluding those probability distributions that do not agree with the macroscopic constraints. If the macroscopic constraints that interest us are based on trait differences that bias demographic processes of immigration, survival and reproduction in a local community, then our prior should represent the theoretical relative abundances (i.e. probabilities) that would occur if traits had no effect on such demographic processes. If we believe that selection on traits is the *only* process that could cause one species to be more abundant than another then this prior would indeed be a maximally uninformative one, as used in Chapters 4 and 6. This is because, if traits are the only processes that could cause one species to be more abundant than another, then once we have the information about these traits in the form of constraint equations we will have explained everything. There will be no other information available to include in the prior and so it will be maximally uninformative. What if we believe that there *are* factors that affect differences in abundance but that are independent of trait differences? Then we must either include macroscopic properties based on such non-trait factors as further constraints, or else modify the prior to implicitly include these non-trait factors. If we include the information on non-trait factors in the prior then this prior is no longer "maximally uninformative". In one sense this is not ideal because, by definition, we cannot claim to have explained information that is hidden in the prior rather than explicitly stated as constraints. In another sense this is exactly what we should do if we are not interested in explaining such non-trait-based processes.

As an example of a non-trait-based constraint on relative abundance, consider the forests of northeastern North America and the forests of northern Europe. Anyone who visits these two forests is struck by the many similarities in the vegetation and in the environment. There are, of course, obvious differences as well. For instance, the observed relative abundance of Sugar Maples in a Dutch forest would be zero almost everywhere in the Netherlands yet this species is common in the forests in my area. This is not because the functional traits of Sugar Maples prevent them from surviving in Holland but allow them to dominate forests in southern Canada. If someone were to plant Sugar Maple trees in a Dutch forest (I hope they don't . . .) then these trees would probably grow and reproduce just fine. The reason for a lack of Sugar Maples in northern Europe is due to phylogenetic history, not functional traits. There is a very powerful present-day constraint at work as well (the Atlantic Ocean) but we would not want to have to include this in our constraint equations. On reflection, there are many historical and geographical processes at many different temporal and spatial scales – and quite independent of functional traits – that affect the distribution of species. If we choose a maximally uninformative prior then we must include these processes if they are important in predicting the structure of the vegetation from a species pool or else accept a poor predictive performance. If we use a small species pool from a small geographical area, as we did in Chapter 6, then perhaps we can ignore such processes, both in our prior and in our constraints, but as our species pool increases in size and in geographical extent then we will either have to modify our prior or else include these non-trait-based processes in our constraints. Because we are ecologists rather than biogeographers, we will modify our prior.

Such non-trait-based processes do not only exist at very large geographical scales. Consider some larger landscape containing the species in our species pool and our local community within this larger landscape. The local community should be small enough[1] that there does not exist strong environmental gradients within it since our process of trait-based environmental filtering (if it exists) occurs primarily within this local community. The larger landscape will normally include many different habitats. The processes affecting immigration of new propagules from the larger landscape into the local community will certainly affect the relative abundance of species in this local community, and some of these processes of immigration will depend on functional traits, but many of these processes will not. For instance, imagine two grassland species (A and B) that occur in the larger landscape and who

---

[1] Perhaps just a few square meters if dealing with herbaceous vegetation.

are very similar in their functional traits except that species A is often planted in pastures while species B is not. If the landscape has many areas that are former pastures then species A will be common and species B will be rare in the landscape because of past land use. If so then species A will contribute more immigrants to our local community than species B quite independently of any functional traits. We would expect to find a higher relative abundance of species A even though this higher abundance is not due to functional traits.[2] How can we deal with this in the context of the Maximum Entropy Formalism?

We can start by making the same assumptions as made in "neutral" models of community assembly (Bell 2000, Hubbell 2001). Let's first assume that every individual of every species in the species pool has exactly the same demographic probabilities of immigration into the local community, and of survival and reproduction once present. If this is true then there can be no trait-based environmental filtering in the local community; this was explained in more detail in Chapter 5. If this is true then species with more individuals will produce more immigrants and the proportion of all immigrants contributed by a given species will be closely related to the relative abundance of the species in the landscape. The probability of a *species* immigrating into the local community will be determined entirely by the relative abundance of the species[3] in the larger landscape. Species that have more individuals in the landscape, for whatever reason, will contribute more immigrants to the local community since each individual has the same chances of immigrating. Since all individuals in the local community have the same probabilities of survival and reproduction then this neutral scenario predicts that the relative abundance of species in the local community will be, on *average*, the same as their relative abundances in the larger landscape. Certainly, random demographic fluctuations in the local community will cause some species to increase and other to decrease but, precisely because they are random, they will be unpredictable. This was explained in more detail in Chapter 5. Given this assumption then the prior probabilities, before we begin to consider possible trait-based biases occurring in the local community, will be the relative abundances of each species in the larger landscape. This defines our "neutral" prior since it is our expectation in the absence of any trait-based environmental filtering of the species pool in the local community. *Why* some species will be more common in the larger landscape is left unexplained – information about such non-trait-based processes at the landscape level is implicitly included

---

[2] Of course, it would be due to at least one "trait": it is preferred by farmers.
[3] More precisely, by the relative abundance of the propagules produced by the species.

in the prior. If community assembly is not neutral in the local community then the actual distribution will be pushed away from this prior distribution due to environmental filtering of the traits possessed by the different species, but such trait-based constraints can be introduced into the model explicitly through the community-aggregated traits. By choosing a neutral prior we are claiming that the relative abundance in the local community of the species in our species pool will be as close as possible to our neutral expectation (the prior) while respecting the community-aggregated traits of the local community that reflects the action of natural selection.

But, you might be thinking, this idea of a neutral prior is too ambiguous to be applicable. You might be right. After all, if I chose to define the landscape as the area within a 1 kilometer radius of the local community while you chose a radius of 10 kilometers, then we will very likely arrive at a different neutral prior. Perhaps a third person might decide to weight the abundances of each species in the landscape by the distance from the local community? If so, how should the weighting be done; inversely proportional to distance or inversely proportional to the square of the distance? These are all good questions and my only answer is: try them all! I don't know of any good theoretical reason to prefer one to the other and so an empirical approach is needed. However, if I could hazard a guess, I suspect that the different ways of defining the neutral prior will not make a large difference except when non-trait-based constraints are particularly strong. This is because the importance of the prior decreases as the strength of the macroscopic constraints increase. I could even imagine using a very imprecise neutral prior. For instance, one could simply divide the landscape into a few basic habitat types identifiable by aerial photographs or remote sensing (wetlands, forested areas, upland grasslands, cultivated land, heavily disturbed areas, etc.). One would then ask local botanists to rank each species in the species pool into a few abundance categories (dominant, common subdominant, rare, absent) for each habitat type. Multiplying these abundance categories by the relative abundance of the habitats, and then standardizing, would produce a prior that roughly reflects the dominant, subdominant and rare species in the larger landscape.

Needless to say, this is pure speculation. The best way to avoid speculating is to assemble real data and put the ideas of this book to the test. Now it's your turn.

References

Ackerly, D. D., and W. K. Cornwell. (2007). A trait-based approach to community assembly: partitioning of species trait values into within- and among-community components. *Ecology Letters* **10**:135–145.

Ackerly, D. D., S. A. Dudley, S. E. Sultan *et al.* (2000). The evolution of plant ecophysiological traits: Recent advances and future directions. *BioScience* **50**:979–995.

Ackerly, D. D., C. A. Knight, S. B. Weiss, K. Barton, and K. P. Starmer. (2002). Leaf size, specific leaf area and microhabitat distribution of chaparral woody plants: contrasting patterns in species level and community level analyses. *Oecologia* **130**:449–457.

Aerts, R. (1990). Nutrient use efficiency in evergreen and deciduous species from heathlands. *Oecologia* **84**:391–397.

Akaike, H. (1973). Information theory and an extension of the maximum likelihood principle. *In* Petrov, B. N., and F. Csaki, editors. *Proceedings of the 2nd International Symposium on Information Theory*. Akademiai Kiado, Budapest.

Allen, A. P., B. Li, and E. L. Charnov. (2001). Population fluctuations, power laws and mixtures of lognormal distributions. *Ecology Letters* **4**:1–3.

Arnold, S. J. (1983). Morphology, performance and fitness. *American Zoologist* **23**:347–361.

Austin, M. P. (1985). Continuum concept, ordination methods, and niche theory. *Annual Review of Ecology and Systematics* **16**:39–61.

Austin, M. P., and T. M. Smith. (1989). A new model for the continuum concept. *Vegetatio* **83**:35–47.

Austin, M. P., R. B. Cunningham, and P. M. Fleming. (1984). New approaches to direct gradient analysis using environmental scalars and statistical curve-fitting procedures. *Vegetatio* **55**:11–27.

Baker, R., and G. Christakos. (2007). Revisiting prior distributions, Part I: Priors based on a physical invariance principle. *Stochastic Environmental Research and Risk Assessment.* **21**:427–434.

Baraloto, C., D. E. Goldberg, and D. Bonal. (2005). Performance trade-offs among tropical tree seedlings in contrasting microhabitats. *Ecology* **86**:2461–2472.

Bazzaz, F. A. (1979). The physiological ecology of plant succession. *Annual Review of Ecology and Systematics* **10**:351–371.

Bazzaz, F. A., and T. W. Sipe. (1987). Physiological ecology, disturbance, and ecosystem recovery. *Ecological Studies* **61**:203–227.

Becks, L., F. M. Hilker, H. Malchow, K. Jurgens, and H. Arndt. (2005). Experimental demonstration of chaos in a microbial food web. *Nature* **435**:1226–1229.

Bell, G. (1996). *The Basics of Selection*. Chapman and Hall, New York.

Bell, G. (2000). The distribution of abundance in neutral communities. *American Naturalist* **155**:606–617.

Bell, G., and M. J. Lechowicz. (1994). The flip side: manifestations of how plants perceive patchiness at different scales. *In* Caldwell, M. M., and R. W. Pearcy, editors. *Exploitation of Environmental Heterogeneity by Plants. Ecophysiological processes above and below ground*. Academic Press, New York, pp. 391–414.

Bell, G., M. J. Lechowicz, and M. J. Waterway. (2006). The comparative evidence relating to functional and neutral interpretations of biological communities. *Ecology* **87**:1378–1386.

Beninca, E., J. Huisman, R. Heerkloss *et al.* (2008). Chaos in a long-term experiment with a plankton community. *Nature* **451**:822–827.

Beveridge, W. I. B. (1957). *The Art of Scientific Investigation*, 3rd edition. Random House, New York.

Blackman, G. E., and G. L. Wilson. (1951). Physiological and ecological studies in the analysis of plant environment. VII. An analysis of the differential effects of light intensity on the net assimilation rate, leaf-area ratio, and relative growth rate of different species. *Annals of Botany* **15**:373–408.

Booth, B. D., and D. W. Larson. (1999). Impact of language, history, and choice of system on the study of assembly rules. *In* Weiher, E., and P. Keddy, editors. *Ecological Assembly Rules. Perspectives, advances, retreats*. Cambridge University Press, Cambridge, pp. 206–229.

Borcard, D., P. Legendre, and P. Drapeau. (1992). Partialling out the spatial component of ecological variation. *Ecology* **73**:1045–1055.

Bouma, J. (1989). Using soil survey data for quantitative land evaluation. *In* Stewart, B. A., editor. *Advances in Soil Science*. Springer-Verlag, New York, pp. 225–239.

Box, E. O. (1981). *Macroclimate and Plant Forms: An introduction to predictive modelling in phytogeography*. Kluwer, The Hague.

Box, E. O. (1995). Factors determining distributions of tree species and plant functional types. *Vegetatio* **121**:101–116.

Box, E. O. (1996). Plant functional types and climate at the global scale. *Journal of Vegetation Science* **7**:309–320.

Braak, C. J. F. t. (1994). Canonical community ordination. Part I: basic theory and linear methods. *Ecoscience* **1**:127–140.

Bradshaw, A. D. (1987). Functional ecology = comparative ecology? *Functional Ecology* **1**:71–72.

Braun-Blanquet, J. (1919). Notions d'élément et de territoire phytogéographiques. *Archives des Sciences Physiques et Naturelles* **1**:497–512.

Burnham, K. P., and D. R. Anderson. (2004). Multimodel inference – understanding AIC and BIC in model selection. *Sociological Methods & Research* **33**:261–304.

Caccianiga, M., A. Luzzaro, S. Pierce, R. M. Ceriani, and B. Cerabolini. (2006). The functional basis of a primary succession resolved by CSR classification. *Oikos* **112**:10–20.

Calow, P. (1987). Towards a definition of functional ecology. *Functional Ecology* **1**:57–61.

Cerabolini, B., R. M. Ceriani, M. Caccianiga, R. D. Andreis, and B. Raimondi. (2003). Seed size, shape and persistence in soil: a test on Italian flora from Alps to Mediterranean coasts. *Seed Science Research* **13**:75–85.

Chapman, R. (1931). *Animal Ecology*. McGraw-Hill, New York.

Cingolani, A. M., M. Cabido, D. E. Gurvich, D. Renison, and S. Diaz. (2007). Filtering processes in the assembly of plant communities: Are species presence and abundance driven by the same traits? *Journal of Vegetation Science* **18**:911–920.

Clatworthy, J. N., and J. L. Harper. (1962). The comparative biology of closely related species living in the same area. V. Inter- and intraspecific interference within cultures of *Lemna* spp. and *Salvinia natans*. *Journal of Experimental Botany* **13**:307–324.

Clements, F. E. (1916). *Plant Succession: An analysis of the development of vegetation*. Carnegie Institution of Washington, Washington.

Clements, F. E. (1936). Nature and structure of the climax. *Journal of Ecology* **24**:252–284.

Conner, E. F., and D. Simberloff. (1979). The assembly of species communities: chance or competition? *Ecology*:1132–1140.

Cooper, W. S. (1926). The fundamentals of vegetational change. *Ecology* **7**:391–413.

Corbet, A. S. (1941). The distribution of butterflies in the Malay Peninsula (Lepid.). *Proceedings of the Royal Entomological Society, Series A* **16**:101–116.

Cornelissen, J. H. C. (1996). An experimental comparison of leaf decomposition rates in a wide range of temperate plant species and types. *Journal of Ecology* **84**:573–582.

Cornelissen, J. H. C., and K. Thompson. (1997). Functional leaf attributes predict litter decomposition rate in herbaceous plants. *New Phytologist* **135**:109–114.

Cornelissen, J. H. C., H. M. Quested, D. Gwynn-Jones *et al.* (2004). Leaf digestibility and litter decomposability are related in a wide range of subarctic plant species and types. *Functional Ecology* **18**:779–786.

Cornelissen, J. H. C., S. I. Lang, N. A. Soudzilovskaia, and H. J. During. (2007). Comparative cryptogam ecology: a review of bryophyte and lichen traits that drive biogeography. *Annals of Botany* **99**:987–1001.

Cornwell, W. K., D. W. Schwilk, and D. D. Ackerly. (2006). A trait-based test for habitat filtering: Convex hull volume. *Ecology* **87**:1465–1471.

Costantino, R. F., R. A. Desharnais, J. M. Cushing *et al.* (2005). Nonlinear stochastic population dynamics: The Flour Beetle *Tribolium* as an effective tool of discovery. *In* Advances in Ecological Research, Vol. 37: *Population Dynamics and Laboratory Ecology*. Academic Press, London, pp. 101–141.

Craine, J. M., D. Tilman, D. Wedin, P. Reich, and M. Tjoelker. (2002). Functional traits, productivity and effects on nitrogen cycling of 33 grassland species. *Functional Ecology* **16**:563–574.

Cramér, H. (1946). *Mathematical Methods of Statistics*. Princeton University Press, Princeton.

Dale, M. R. T., and M. J. Fortin. (2002). Spatial autocorrelation and statistical tests in ecology. *Ecoscience* **9**:162–167.

Darwin, C. (1859). *On the Origin of Species*. John Murray, London.

Darwin, C., and A. R. Wallace. (1858). On the tendency of species to form varieties; and on the perpetuation of varieties and species by natural means of selection. *Proceedings of the Linnean Society London, Zoology* **3**:45–62.

De Bello, F., J. Leps, and M. T. Sebastia. (2005). Predictive value of plant traits to grazing along a climatic gradient in the Mediterranean. *Journal of Applied Ecology* **42**:824–833.

Dear, P. (2007). *The Intelligibility of Nature. How science makes sense of the world*. University of Chicago Press, Chicago.

Debussche, M., J. Lepart, and A. Dervieux. (1999). Mediterranean landscape changes: evidence from old postcards. *Global Ecology and Biogeography* **8**:3–15.

Della Pietra, S., V. Della Pietra, and J. Lafferty. (1997). Inducing features of random fields. *IEEE Transactions Pattern Analysis and Machine Intelligence* **19**:1–13.

Dennett, D. C. (1995). *Darwin's Dangerous Idea: Evolution and the meanings of life*. Simon and Schuster.

Dewar, R. (2003). Information theory explanation of the fluctuation theorem, maximum entropy production and self-organized criticality in non-equilibrium stationary states. *Journal of Physics A – Mathematical and General* **36**:631–641.

Dewar, R. (2004). Maximum entropy production and non-equilibrium statistical mechanics. *In* Kliedon, A., and R. D. Lorenz, editors. *Non-equilibrium Thermodynamics and the Production of Entropy: Life, earth and beyond*. Springer-Verlag, Berlin, pp. 41–56.

Dewar, R. (2005). Maximum entropy production and the fluctuation theorem. *Journal of Physics A – Mathematical and General* **38**:L371–L381.

Dewar, R. C., and A. Porté. (2008). Statistical mechanics unifies different ecological patterns. *Journal of Theoretical Biology* **251**:389–403.

Diamond, J. M. (1975). Assembly of species communities. *In* Cody, M. L., and J. M. Diamond, editors. *Ecology and Evolution of Communities*. Belknap Press, Cambridge, pp. 342–444.

Diaz, S., and M. Cabido. (2001). Vive la difference: plant functional diversity matters to ecosystem processes. *Trends in Ecology & Evolution* **16**:646–655.

Diaz, S., M. Cabido, and F. Casanoves. (1998). Plant functional traits and environmental filters at a regional scale. *Journal of Vegetation Science* **9**:113–122.

Diaz, S., J. G. Hodgson, K. Thompson *et al.* (2004). The plant traits that drive ecosystems: Evidence from three continents. *Journal of Vegetation Science* **15**:295–304.

Diekmann, M. (2003). Species indicator values as an important tool in applied plant ecology – a review. *Basic and Applied Ecology* **4**:493–506.

Diemer, M. (1998). Life span and dynamics of leaves of herbaceous perennials in high-elevation environments: 'news from the elephant's leg'. *Functional Ecology* **12**:413–425.

Dormann, C. F., and S. H. Roxburgh. (2005). Experimental evidence rejects pairwise modelling approach to coexistence in plant communities. *Proceedings of the Royal Society of London* B, **272**:1279–1285.

DuRietz, G. E. (1931). Life-forms of terrestrial flowering plants I. *Acta Phytogeog. Suecica* **3**:1–95.

Dybzinski, R., and D. Tilman. (2007). Resource use patterns predict long-term outcomes of plant competition for nutrients and light. *American Naturalist* **170**:305–318.

Dykserhuis, E. J. (1949). Condition and management of range land, based on quantitative ecology. *Journal of Range Management* **2**:104–115.

Egolf, D. A. (2001). Equilibrium regained: from nonequilibrium chaos to statistical mechanics. *Science* **287**:101–104.

Elger, A., and N. J. Willby. (2003). Leaf dry matter content as an integrative expression of plant palatability: the case of freshwater macrophytes. *Functional Ecology* **17**:58–65.

Ellenberg, H. (1978). *Vegetation Mitteleuropas mit den Alpen in ökologischer Sicht*. Verlag.

Escarré, J., C. Houssard, M. Debussche, and J. Lepart. (1983). Évolution de la végétation et du sol après abandon cultural en région méditerranéene: étude de succession dans les garrigues du Montpelliérais (France). *Acta Oecologica* **4**:221–239.

Etienne, R. S. (2005). A new sampling formula for neutral biodiversity. *Ecology Letters* **8**:253–260.

Etienne, R. S. (2007). A neutral sampling formula for multiple samples and an 'exact' test of neutrality. *Ecology Letters* **10**:608–618.

Falconer, D. S., and T. F. C. Mackay. (1996). *Introduction to Quantitative Genetics*. Prentice Hall, London.

Fenner, M., and K. Kitajima. (1999). Seed and seedling ecology. *In* Pugnaire, F., and F. Valladares, editors. *Handbook of Functional Plant Ecology*. Marcel Dekker, New York, pp. 589–621.

Fenner, M., and T. Lee. (1989). Growth of seedlings of pasture grasses and legumes deprived of single mineral nutrients. *Journal of Applied Ecology* **26**:223–232.

Fisher, R. A. (1925). *Statistical Methods for Research Workers*, first edition. Oliver & Boyd, Edinburgh.

Fisher, R. A. (1928a). The general sampling distribution of the multiple correlation coefficient. *Proceedings of the Royal Society of London* A, **121**:654–673.

Fisher, R. A. (1928b). Moments and product moments of sampling distributions. *Proceedings of the London Mathematical Society* **30**:199–238.

Fisher, R. A., A. S. Corbet, and C. B. Williams. (1943). The relation between the number of species and the number of individuals in a random sample from an animal population. *Journal of Animal Ecology* **12**:42–58.

Fonseca, C. R., J. M. Overton, B. Collins, and M. Westoby. (2000). Shifts in trait-combinations along rainfall and phosphorus gradients. *Journal of Ecology* **88**:964–977.

Foster, S. A. (1986). On the adaptive value of large seeds for tropical moist forest trees: a review and synthesis. *The Botanical Review* **52**:260–297.

Fougere, P. F. (1988). Maximum entropy calculations on a discrete probability space. *In* Erickson, G. J., and C. R. Smith, editors. *Maximum-Entropy and Bayesian Methods in Science and Engineering*. Kluwer, Dordrecht, pp. 205–234.

Franzaring, J., A. Fangmeier, and R. Hunt. (2007). On the consistencies between CSR plant strategies and Ellenberg ecological indicator values. *Journal of Applied Botany and Food Quality-Angewandte Botanik* **81**:86–94.

Fukami, T., T. M. Bezemer, S. R. Mortimer, and W. H van der Putten. (2005). Species divergence and trait convergence in experimental plant community assembly. *Ecology Letters* **8**:1283–1290.

Garnier, E. (1991). Resource capture, biomass allocation and growth in herbaceous plants. *Trends in Ecology & Evolution* **6**:126–131.

Garnier, E., G. Laurent, A. Bellmann *et al.* (2001). Consistency of species ranking based on functional leaf traits. *New Phytologist* **152**:69–83.

Garnier, E., J. Cortez, G. Billes *et al.* (2004). Plant functional markers capture ecosystem properties during secondary succession. *Ecology* **85**:2630–2637.

Garnier, E., S. Lavorel, P. Ansquer *et al.* (2007). Assessing the effects of land-use change on plant traits, communities and ecosystem functioning in grasslands: A standardized methodology and lessons from an application to 11 European sites. *Annals of Botany* **99**:967–985.

Gassmann, F., F. Klotzli, and G. R. Walther. (2005). Vegetation change shows generic features of non-linear dynamics. *Journal of Vegetation Science* **16**:703–712.

Gaucherand, S., and S. Lavorel. (2007). New method for rapid assessment of the functional composition of herbaceous plant communities. *Austral Ecology* **32**:927–936.

Gaudet, C. L., and P. A. Keddy. (1988). A comparative approach to predicting competitive ability from plant traits. *Nature* **334**:242–243.

Gause, G. F. (1934). *The Struggle for Existence*. Williams and Wilkins, Baltimore.

Gilpin, M. E. (1979). Spiral chaos in a predator-prey model. *American Naturalist* **113**:306–308.

Gilpin, M. E., M. P. Carpenter, and M. J. Pomerantz. (1986). The assembly of a laboratory community: multispecies competition in *Drosophila*. *In* Diamond, J. M., and T. Case, editors. *Community Ecology*. Harper & Row, Cambridge.

Givnish, T. J. (2002). Adaptive significance of evergreen vs. deciduous leaves: solving the triple paradox. *Silva Fennica* **36**:703–743.

Gleason, H. A. (1926). The individualistic concept of the plant association. *Bulletin of the Torry Club.* **53**:7–26.

Gleick, J. (1987). *Chaos: Making a new science*. Penguin, New York.

Goldsmith, F. B. (1978). Interaction (competition) studies as a step towards the synthesis of sea-cliff vegetation. *Journal of Ecology* **66**:921–931.

Grace, J. B. (1999). The factors controlling species density on herbaceous plant communities: an assessment. *Perspectives in Plant Ecology, Evolution and Systematics.* **2**:1–28.

Grace, J. B., and B. H. Pugesek. (1997). A structural equation model of plant species richness and its application to coastal wetland. *American Naturalist* **149**:436–460.

Grace, J. B., and B. H. Pugesek. (1998). On the use of path analysis and related procedures for the investigation of ecological problems. *American Naturalist* **152**:151–159.

Grace, J. B., L. Allain, and C. Allen. (1999). Plant species density in a coastal tallgrass prairie: the importance of environmental effects. In review.

Grime, J. P. (1974). Vegetation classification by reference to strategies. *Nature* **250**:26–31.

Grime, J. P. (1977). Evidence for the existence of three primary strategies in plants and its relevence to ecological and evolutionary theory. *American Naturalist* **111**:1169–1194.

Grime, J. P. (1979). *Plant Strategies and Vegetation Processes*. John Wiley & Sons, New York.

Grime, J. P. (1998). Benefits of plant diversity to ecosystems: immediate, filter and founder effects. *Journal of Ecology* **86**:902–910.

Grime, J. P. (2001). *Plant Strategies, Vegetation Processes, and Ecosystem Properties*, 2nd edition. John Wiley & Sons, New York.

Grime, J. P. (2007). *Comparative Plant Ecology*. Castlepoint Press, Colvend.

Grime, J. P., and R. Hunt. (1975). Relative growth rate: its range and adaptive significance in a local flora. *Journal of Ecology* **63**:393–422.

Haegeman, B., and M. Loreau. (2008). Limitations of entropy maximization in ecology. *Oikos* **117**:1700–1710.

Halloy, S. R. P., and A. F. Mark. (1996). Comparative leaf morphology spectra of plant communities in New Zealand, the Andes and the European Alps. *Journal of the Royal Society of New Zealand* **26**:41–78.

Harper, J. L. (1982). After description. *In* Newman, E. I., editor. *The Plant Community as a Working Mechanism*. Blackwell Scientific, Oxford, pp. 11–25.

Hasti, T. J., and R. J. Tibshirani. 1990. *Generalized Additive Models*. Chapman and Hall, London.

Hastings, A., and T. Powell. (1991). Chaos in a 3-Species Food-Chain. *Ecology* **72**:896–903.

Hempel, C. G. (1965). *Aspects of Scientific Explanation and Other Essays in the Philosophy of Science*. Free Press., New York.

Hendry, G. A. F., and J. P. Grime. (1993). *Methods in Comparative Plant Ecology*. Chapman & Hall, London.

Herschel, J. F. W. (1872). *Physical Geography of the Globe*. A. and C. Black, London.

Hewitt, N. (1998). Seed size and shade-tolerance – A comparative analysis of North American temperate trees. *Oecologia* **114**:432–440.

Hodgson, J. G., P. J. Wilson, R. Hunt, J. P. Grime, and K. Thompson. (1999). Allocating C-S-R plant functional types: a soft approach to a hard problem. *Oikos* **85**:282–294.

Holdridge, L. R. (1947). Determination of world plant formations from simple climatic data. *Science* **105**:367–368.

Hox, J. J. (2002). *Multilevel Analysis: Techniques and Applications*. Lawrence Erlbaum, Mahwah, NJ.

Hubbell, S. P. (2001). *The Unified Neutral Theory of Biodiversity and Biogeography*. Princeton University Press, Princeton.

Hubbell, S. P. (2006). Neutral theory and the evolution of ecological equivalence. *Ecology* **87**:1387–1398.

Huisman, J., N. N. P. Thi, D. M. Karl, and B. Sommeijer. (2006). Reduced mixing generates oscillations and chaos in the oceanic deep chlorophyll maximum. *Nature* **439**:322–325.

Huisman, J., and F. J. Weissing. (1999). Biodiversity of plankton by species oscillations and chaos. *Nature* **402**:407–410.

Hutchinson, G. E. (1959). Homage to Santa Rosalia or why are there so many kinds of animals? *American Naturalist* **53**:145–159.

Hutchinson, G. E. (1978). *An Introduction to Population Ecology*. Yale University Press, New Haven.

Jain, R., and S. Ramakumar. (1999). Stochastic dynamics modeling of the protein sequence length distribution in genomes: implications for microbial evolution. *Physica A* **273**:476–485.

Jaynes, E. T. (1957a). Information theory and statistical mechanics I. *The Physical Review* **106**:620–630.

Jaynes, E. T. (1957b). Information theory and statistical mechanics II. *The Physical Review* **108**:171–190.

Jaynes, E. T. (1971). Violations of Boltzmann's H theorem in real gases. *Physical Reviews* **A4**:747–751.

Jaynes, E. T. (1983). *Papers on Probability, Statistics, and Statistical Physics*. D. Reidel, Dordrecht.

Jaynes, E. T. (2003). *Probability Theory. The Logic of Science*. Cambridge University Press, Cambridge.

Jeffreys, H. (1939). *Theory of Probability*. Clarendon Press, Oxford.

Kazakou, E., D. Vile, B. Shipley, C. Gallet, and E. Garnier. (2006). Co-variations in litter decomposition, leaf traits and plant growth in species from a Mediterranean old-field succession. *Functional Ecology* **20**:21–30.

Keddy, P. A. (1982). Population ecology on an environmental gradient.: *Cakile edentula* on a sand dune. *Oecologia* **52**:348–355.

Keddy, P. A. (1984). Plant zonation on lakeshores in Nova Scotia: a test of the resource specialization hypthesis. *Journal of Ecology* **72**:797–808.

Keddy, P. A. (1990). The use of functional as opposed to phylogenetic systematics: a first step in predictive community ecology. *In* Kawano, S., editor. *Biological Approaches and Evolutionary Trends in Plants*. Harcourt Brace Jovanovich, London, pp. 387–406.

Keddy, P. A. (1992). Assembly and response rules: two goals for predictive community ecology. *Journal of Vegetation Science* **3**:157–164.

Keddy, P. A. (2001). *Competition*, second edition. Cambridge University Press, Cambridge.

Keddy, P. A., and B. Shipley. (1989). Competitive hierarchies in herbaceous plant communities. *Oikos* **54**:234–241.

Kenkel, N. C., P. Juhasz-Nagy, and J. Podani. (1989). On sampling procedures in population and community ecology. *Vegetatio* **83**:195–207.

Khurana, E., and J. S. Singh. (2000). Influence of seed size on seedling growth of *Albizia procera* under different soil water levels. *Annals of Botany* **86**:1185–1192.

Kikuzawa, K. (1991). A cost-benefit analysis of leaf habit and leaf longevity of trees and their geographical pattern. *American Naturalist* **138**:1250–1260.

Kikuzawa, K. (1995). The basis for variation in leaf longevity of plants. *Vegetatio* **121**:89–100.

Klebanoff, A., and A. Hastings. (1994). Chaos in 3 species food-chains. *Journal of Mathematical Biology* **32**:427–451.

Kleyer, M. (1995). *Biological Traits of Vascular Plants. A database. Albeitsberichte Inst. f. Landschaftsplanung u. Okologie*. University of Stuttgart, Stuttgart.

Knevel, I. C., R. M. Bekker, J. P. Bakker, and M. Kleyer. (2003). Life-history traits of the northwest European flora: The LEDA database. *Journal of Vegetation Science* **14**:611–614.

Knight, D. H. (1965). A gradient analysis of Wisconsin prairie vegetation on the basis of plant structure and function. *Ecology* **46**:744–747.

Kolmogoroff, A. N. (1956). *Foundations of the Theory of Probability*, second edition. Chelsea, New York.

Kuhn, I., W. Durka, and S. Klotz. (2004). BiolFlor – a new plant-trait database as a tool for plant invasion ecology. *Diversity and Distributions* **10**:363–365.

Kullback, S. (1959). *Information Theory and Statistics*. Wiley, New York.

Kullback, S., and R. A. Leibler. (1951). On information and sufficiency. *Annals of Mathematical Statistics* **22**:79–86.

Lambers, H., and H. Poorter. (1992). Inherent variation in growth rate between higher plants: A search for physiological causes and ecological consequences. *Advances in Ecological Research* **23**:187–261.

Lambers, H., F. S. Chapin, and T. L. Pons. (1998). *Plant Physiological Ecology*. Springer, New York.

Lande, R., and S. J. Arnold. (1983). The measurement of selection on correlated characters. *Evolution* **37**:1210–1226.

Lange, O. L., P. S. Nobel, C. B. Osmond, and H. Zeigler. (1983). *Physiological Plant Ecology*. Springer-Verlag, Berlin.

Larcher, W. (2001). *Physiological Plant Ecology*, fourth edition. Springer, Berlin.

Larson, D. W. (2000). *Cliff Ecology: Pattern and process in cliff ecosystems*. Cambridge University Press, Cambridge.

Lavorel, S., and E. Garnier. (2002). Predicting changes in community composition and ecosystem functioning from plant traits: revisiting the Holy Grail. *Functional Ecology* **16**:545–556.

Lechowicz, M. J., and P. A. Blais. (1988). Assessing the contributions of multiple interacting traits to plant reproductive success: environmental dependence. *Journal of Evolutionary Biology* **1**:255–273.

Legendre, P., and M. J. Anderson. (1999). Distance-based redundancy analysis: testing multispecies responses in multifactorial ecological experiments. *Ecological Monographs* **69**:1–24.

Legendre, P., and L. Legendre. (1998). *Numerical Ecology*, second edition. Elsevier, Amsterdam.

Leps, J., and P. Smilauer. (2003). *Multivariate Analysis of Ecological Data using CANOCO*. Cambridge University Press, Cambridge.

Levins, R. (1966). The strategy of model building in population biology. *American Scientist* **54**:421–431.

Li, W. (1996). Random texts exhibit Zipf's-law-like word frequency distribution. *IEEE Transactions on Information Theory* **38**:1842–1845.

Limpert, E., W. A. Stahel, and M. Abbt. (2001). Log-normal distributions across the sciences: keys and clues. *BioScience* **51**:341–352.

Lososova, Z., M. Chytry, I. Kuhn *et al.* (2006). Patterns of plant traits in annual vegetation of man-made habitats in central Europe. *Perspectives in Plant Ecology Evolution and Systematics* **8**:69–81.

Lotka, A. J. (1925). *Elements of Physical Biology*. Williams & Wilkins, Baltimore.

MacArthur, R. H. (1972). *Geographical Ecology: Patterns in the Distribution of Species*. Harper and Row, New York.

MacArthur, R. H., and R. Levins. (1967). The limiting similarity, convergence, and divergence of coexisting species. *American Naturalist* **101**:377–385.

MacLeod, J. (1894). Over de bevruchting der bloemen in het Kempisch gedeelte van Vlaanderen. Deel II. *Botanische Jaarboek* **6**:119–511.

Mandelbrot, B. (1997). *Fractals and Scaling in Finance*. Springer-Verlag, New York.

Manly, B. F. J. (1997). *Randomization, Bootstrap and Monte Carlo Methods in Biology*, second edition. Chapman and Hall, London.

Marks, C. O., and M. J. Lechowicz. (2006a). Alternative designs and the evolution of functional diversity. *American Naturalist* **167**:55–66.

Marks, C. O., and M. J. Lechowicz. (2006b). A holistic tree seedling model for the investigation of functional trait diversity. *Ecological Modelling* **193**:141–181.

Marks, C. O., and H. C. Muller-Landau. (2007). Comment on "From plant traits to plant communities: A statistical mechanistic approach to biodiversity". *Science* **316**:1425.

May, R. A. (1974). Biological populations with nonoverlapping generations: stable points, stable cycles, and chaos. *Science* **186**:645–647.

May, R. M. (1976). Simple mathematical models with very complicated dynamics. *Nature* **261**:459–467.

McGill, B. J., B. A. Maurer, and M. D. Weiser. (2006). Empirical evaluation of neutral theory. *Ecology* **87**:1411–1423.

McGill, B. J., R. S. Etienne, J. S. Gray et al. (2007). Species abundance distributions: moving beyond single prediction theories to integration within an ecological framework. *Ecology Letters* **10**:995–1015.

Meziane, D. (1998). *Étude de la variation interspécifique de la vitesse spécifique de croissance et modélisation de l'effet des attributs morphologiques, physiologiques et d'allocation de biomasse*. Ph.D. Université de Sherbrooke, Sherbrooke.

Meziane, D., and B. Shipley. (1999). Interacting determinants of specific leaf area in 22 herbaceous species: effects of irradiance and nutrient availability. *Plant, Cell and Environment* **22**:447–459.

Miller, T. E., J. H. Burns, P. Munguia et al. (2005). A critical review of twenty years' use of the resource-ratio theory. *American Naturalist* **165**:439–448.

Minchin, P. R. (1987). An evaluation of the relative robustness of techniques for ecological ordination. *Vegetatio* **69**:89–107.

Mitzenmacher, M. (2004). A brief history of generative models for power law and lognormal distributions. *Internet Mathematics* **1**:226–251.

Moles, A. T., D. D. Ackerly, C. O. Webb et al. (2005). Factors that shape seed mass evolution. *Proceedings of the National Academy of Sciences of the United States of America* **102**:10 540–10 544.

Monod, J. (1950). La technique de culture continue: théorie et applications. *Annales de l'Institut Pasteur* **79**:390–410.

Monteith, J. L. (1977). Climate and the efficiency of crop production in Britain. *Philosophical Transactions of the Royal Society of London* B **281**:277–294.

Mosekilde, E. (1996). *Topics in Nonlinear Dynamics. Applications to physics, biology and economic systems.* World Scientific, Singapore.

Motomura, I. (1932). A statistical treatment of associations [in Japanese]. *Japanese Journal of Zoology* **44**:379–383.

Mulaik, S. A. (2001). The curve-fitting problem: An objectivist view. *Philosophy of Science* **68** :218–241.

Munakata, T. (1998). *Fundamentals of the New Artificial Intelligence: Beyond traditional paradigms.* Springer, New York.

Nakajima, T., and A. Higurashi. (1998). A use of two-channel radiances for an aeresol characterization from space. *Geophysical Research Letters* **25**:3815–3818.

Nekola, J. C., and J. H. Brown. (2007). The wealth of species: ecological communities, complex systems and the legacy of Frank Preston. *Ecology Letters* **10**:188–196.

Niinemets, U. (1999). Components of leaf dry mass per area – thickness and density – alter leaf photosynthetic capacity in reverse directions in woody plants. *New Phytologist* **144**:35–47.

Nowak, M. A. (2006). *Evolutionary Dynamics.* Belknap Press, Cambridge.

Odenbaugh, J. (2005). Idealized, inaccurate but successful: A pragmatic approach to evaluating models in theoretical ecology. *Biology and Philosophy* **20**:231–255.

Ozinga, W. A., R. M. Bekker, J. H. J. Schaminee, and J. M. Van Groenendael. (2004). Dispersal potential in plant communities depends on environmental conditions. *Journal of Ecology* **92**:767–777.

Pachepsky, Y., and W. J. Rawls. (2004). *Development of Pedotransfer Functions in Soil Hydrology,* 30. Elsevier, Amsterdam.

Pearcy, R. W., J. Ehleringer, H. A. Mooney, and P. W. Rundel, editors. (1991). *Plant Physiological Ecology.* Chapman and Hall, London.

Pearson, K. (1903). Mathematical contributions to the theory of evolution. XI. On the influence of natural selection on the variability and correlation of organs. *Philosophical Transactions of the Royal Society of London* **A203**:1–66.

Perline, R. (1996). Zipf's law, the central limit theorem, and the random division of the unit interval. *Physical Review E* **54**:220–223.

Peters, R. H. (1991). *A Critique for Ecology.* Cambridge University Press, Cambridge.

Pinheiro, J. C., and D. M. Bates. (2000). *Mixed-effects Models in S and S-PLUS.* Springer, New York.

Poorter, H., and C. Remkes. (1990). Leaf area ratio and net assimilation rate of 24 species differing in relative growth rate. *Oecologia* **83**:553–559.

Popper, K. (1980). *The Logic of Scientific Discovery,* tenth edition. Hutchinson, London.

Preston, F. W. (1948). The commonness, and rarity, of species. *Ecology* **29**:254–283.

Preston, F. W. (1950). Gas laws and wealth laws. *The Scientific Monthly* **71**:309–311.

Preston, F. W. (1962). The canonical distribution of commonness and rarity. *Ecology* **43**:185–215.

Preston, F. W. (1981). Pseudo-lognormal distributions. *Ecology* **62**:355–364.

Pueyo, S., F. He, and T. Zillio. (2007). The maximum entropy formalism and the idiosyncratic theory of biodiversity. *Ecology Letters* **10**:1017–1028.

Pywell, R. F., J. M. Bullock, D. B. Roy *et al.* (2003). Plant traits as predictors of performance in ecological restoration. *Journal of Applied Ecology* **40**:65–77.

Raunkiaer, C. (1934). *The Life Forms of Plants and Statistical Plant Geography, Being the Collected Papers of C. Raunkiaer.* Clarendon Press, Oxford.

Reich, P. B., M. B. Walters, and D. S. Ellsworth. (1992). Leaf life-span in relation to leaf, plant, and stand characteristics among diverse ecosystems. *Ecological Monographs* **62**:365–392.

Reich, P. B., M. B. Walters, and D. S. Ellsworth. (1997). From tropics to tundra: global convergence in plant functioning. *Proceedings of the National Academy of Sciences* **94**:13 730–13 734.

Reich, P. B., M. B. Walters, D. S. Ellsworth *et al.* (1998). Relationships of leaf dark respiration to leaf nitrogen, specific leaf area and leaf life-span: a test across biomes and functional groups. *Oecologia* **114**:471–482.

Ribichich, A. M. (2005). From null community to non-randomly structured actual plant assemblages: parsimony analysis of species co-occurrences. *Ecography* **28**:88–98.

Rissanen, J. (1983). A universal prior for integers and estimation by minimum description length. *Annals of Statistics* **11**:416–431.

Roche, P., N. Diaz-Burlinson, and S. Gachet. (2004). Congruency analysis of species ranking based on leaf traits: which traits are more reliable? *Plant Ecology* **174**:37–48.

Roxburgh, S. H., and K. Mokany. (2007). Comment on "From plant traits to plant communities: A statistical mechanistic approach to biodiversity". *Science* **316**:1425b.

Roxburgh, S. H., and J. B. Wilson. (2000a). Stability and coexistence in a lawn community: experimental assessment of the stability of the actual community. *Oikos* **88**:409–423.

Roxburgh, S. H., and J. B. Wilson. (2000b). Stability and coexistence in a lawn community: mathematical prediction of stability using a community matrix with parameters derived from competition experiments. *Oikos* **88**:395–408.

Schaffers, A. P., and K. V. Sykora. (2000). Reliability of Ellenberg indicator values for moisture, nitrogen and soil reaction: a comparison with field measurements. *Journal of Vegetation Science* **11**:225–244.

Shannon, C. E., and W. Weaver. (1949). *The Mathematical Theory of Communication.* University of Illinois Press, Urbana.

Shipley, B. (1987a). The relationship between dynamic game theory and the Lotka–Volterra competition equations. *Journal of Theoretical Biology* **125**:121–123.

Shipley, J. W. (1987b). *Pattern and mechanism in the emergent macrophyte communities along the Ottawa River (Canada).* Ph.D. University of Ottawa, Ottawa.

Shipley, B. (1993). A null model for competitive hierarchies in competition matrices. *Ecology* **74**:1693–1699.

Shipley, B. (2000a). *Cause and Correlation in Biology: A user's guide to path analysis, structural equations, and causal inference.* Cambridge University Press, Cambridge.

Shipley, B. (2000b). Plasticity in relative growth rate and its components following a change in irradiance. *Plant, Cell and Environment* **23**:1207–1216.

Shipley, B. (2006). Net assimilation rate, specific leaf area and leaf mass ratio: which is most closely correlated with relative growth rate? A meta-analysis. *Functional Ecology* **20**:565–574.

Shipley, B. (2007). Comparative plant ecology as a tool for integrating across scales. *Annals of Botany* **99**:965–966.

Shipley, B. (2009). Limitations of entropy maximization in ecology: a reply to Haegeman and Loreau. *Oikos* **118**:152–159.

Shipley, B., and J. Dion. (1992). The allometry of seed production in herbaceous angiosperms. *American Naturalist* **139**:467–483.

Shipley, B., and R. Hunt. (1996). Regression smoothers for estimating parameters of growth analyses. *Annals of Botany* **76**:569–576.

Shipley, B., and P. A. Keddy. (1987). The individualistic and community-unit concepts as falsifiable hypotheses. *Vegetatio* **69**:47–55.

Shipley, B., P. A. Keddy, D. R. J. Moore, and K. Lemky. (1989). Regeneration and establishment strategies of emergent macrophytes. *Journal of Ecology* **77**:1093–1110.

Shipley, B., P. A. Keddy, and L. P. Lefkovitch. (1991). Mechanisms producing plant zonation along a water depth gradient – a comparison with the exposure gradient. *Canadian Journal of Botany* **69**:1420–1424.

Shipley, B., and R. H. Peters. (1990). A test of the Tilman model of plant strategies: relative growth rate and biomass partitioning. *American Naturalist* **136**:139–153.

Shipley, B., and R. H. Peters. (1991). The Seduction by Mechanism – a Reply to Tilman. *American Naturalist* **138**:1276–1282.

Shipley, B., M. J. Lechowicz, I. Wright, and P. B. Reich. (2006a). Fundamental trade-offs generating the worldwide leaf economics spectrum. *Ecology* **87**:535–541.

Shipley, B., D. Vile, and E. Garnier. (2006b). From plant traits to plant communities: A statistical mechanistic approach to biodiversity. *Science* **314**:812–814.

Shipley, B., D. Vile, and E. Garnier. (2007). Response to comments on "From plant traits to plant communities: A statistical mechanistic approach to biodiversity". *Science* **316**:1425d.

Sole, D., D. Alonso, and A. McKane. (1987). Connectivity and scaling in S-species model ecosystems. *Physica* A **286**:337–344.

Southwood, T. R. E. (1977). Habitat, the templet for ecological strategies. *Journal of Animal Ecology* **46**:337–365.

Steiger, T. L. (1930). Structure of prairie vegetation. *Ecology* **11**:170–238.

Sugihara, G. (1980). Minimal community structure: An explanation of species abundance distributions. *American Naturalist* **116**:770–787.

Tansley, A. G. (1920). The classification of vegetation and the concept of development. *Journal of Ecology* **8**:118–149.

Tansley, A. G. (1935). The use and abuse of vegetational concepts and terms. *Ecology* **16**:284–307.

Taylor Pegg Jr, E. (1994). *A complete list of fair dice*. MSc. University of Colorado, Colorado Springs.

Ter Braak, C. J. F., and I. C. Prentice. (1988). A theory of gradient analysis. *Advances in Ecological Research* **18**:271–317.

Thompson, K. (1987). Seeds and seed banks. *New Phytologist* **106**:23–34.

Thompson, K., J. G. Hodgson, J. P. Grime, *et al.* (1993a). Ellenberg numbers revisited. *Phytocoenologia* **23**:277–289.

Thompson, K., S. R. Band, and J. G. Hodgson. (1993b). Seed size and shape predict persistence in soil. *Functional Ecology* **7**:236–241.

Thompson, K., J. P. Bakker, R. M. Bekker, and J. G. Hodgson. (1998). Ecological correlates of seed persistence in soil in the north-west European flora. *Journal of Ecology* **86**:163–169.

Tilman, D. (1982). *Resource Competition and Community Structure.* Princeton University Press, Princeton.

Tilman, D. (1988). *Plant Strategies and the Dynamics and Structure of Plant Communities.* Princeton University Press, Princeton.

Tilman, D. (2007). Resource competition and plant traits: a response to *Craine et al.* 2005. *Journal of Ecology* **95**:231–234.

van der Valk, A. G. (1981). Succession in wetlands: a Gleasonian approach. *Ecology* **62**:688–696.

van der Valk, A. G. (1988). From community ecology to vegetation management: providing a scientific basis for management. Pages 463–470 *in Transactions of the 53rd North American Wildlife & Natural Resource Conference.*

Van Hulst, R. (1992). From population dynamics to community dynamics: modelling succession as a species replacement process. *In* Glenn-Lewin, D. C., R. K. Peet, T. T. Veblen, editors. *Plant Succession. Theory and Prediction.* Chapman and Hall, London, pp. 188–214.

Vandermeer, J. H. (1969). The competitive structure of communities: an experimental approach with Protozoa. *Ecology* **50**:362–371.

Vendramini, F., S. Diaz, D. E. Gurvich *et al.* (2002). Leaf traits as indicators of resource-use strategy in floras with succulent species. *New Phytologist* **154**:147–157.

Vile, D. (2005). *Significations fonctionnelle et écologique des traits des espèces végétales: exemple dans une successions post-culturale méditerranéenne et généralisations.* Ph.D. Université de Sherbrooke, Sherbrooke.

Vile, D., B. Shipley, and E. Garnier. (2006a). Ecosystem productivity can be predicted from potential relative growth rate and species abundance. *Ecology Letters* **9**:1061–1067.

Vile, D., B. Shipley, and E. Garnier. (2006b). A structural equation model to integrate changes in functional strategies during old-field succession. *Ecology* **87**:504–517.

Violle, C., J. Lecoeur, and M. L. Navas. (2007a). How relevant are instantaneous measurements for assessing resource depletion under plant cover? A test on light and soil water availability in 18 herbaceous communities. *Functional Ecology* **21**:185–190.

Violle, C., M. L. Navas, D. Vile *et al.* (2007b). Let the concept of trait be functional! *Oikos* **116**:882–892.

Volkov, I., J. R. Banavar, S. P. Hubbell, and A. Maritan. (2003). Neutral theory and relative species abundance in ecology. *Nature* **424**:1035–1037.

Volkov, I., J. R. Banavar, F. L. He, S. P. Hubbell, and A. Maritan. (2005). Density dependence explains tree species abundance and diversity in tropical forests. *Nature* **438**:658–661.

Volterra, V. (1926). Variazioni e fluttuazioni del numero d'individui in specie animali conviventi. *Mem. R. Acad. dei Lincei* (Ser. 6) **2**:31–113.

von Mises, R. (1957). *Probability, Statistics and Truth,* second edition. Constable, London.

Warming, E., and M. Vahl. (1909). *Oecology of Plants – an Introduction to the Study of Plant Communities.* Clarendon Press, Oxford.

Weiher, E., and P. A. Keddy. (1995). Assembly rules, null models and trait dispersion. *Oikos* **74**:159–164.

Weiher, E., G. D. P. Clarke, and P. A. Keddy. (1998). Community assembly rules, morphological dispersion, and the coexistence of species. *Oikos* **81**:309–322.

Weiher, E., A. van der Werf, K. Thompson *et al.* (1999). Challenging Theophrastus: A common core list of plant traits for functional ecology. *Journal of Vegetation Science* **10**:609–620.

Weiher, E., S. Forbes, T. Schauwecker, and J. B. Grace. (2004). Multivariate control of plant species richness and community biomass in blackland prairie. *Oikos* **106**:151–157.

Westoby, M. (1998). A leaf-height-seed (LHS) plant ecology strategy scheme. *Plant & Soil* **199**:213–227.

Wheatland, M. S., and P. A. Sturrock. (1996). Avalanche models of solar flares and the distribution of active regions. *The Astrophysical Journal* **471**:1044–1048.

Whittaker, R. H. (1951). A criticism of the plant association and climatic climax concepts. *Northwest Science* **25**:17–31.

Whittaker, R. H. (1956). Vegetation of the Great Smoky Mountains. *Ecological Monographs* **26**:1–80.

Whittaker, R. H. (1965). Dominance and diversity in land plant communities. *Science* **147**:250–260.

Whittaker, R. H. (1967). Gradient analysis of vegetation. *Biological Reviews of the Cambridge Philosophical Society* **49**:207–264.

Whittaker, R. H. (1970). *Communities and Ecosystems*. Macmillan, New York.

Woodward, F. I. (1987). *Climate and Plant Distribution*. Cambridge University Press, Cambridge.

Woodward, F. I. (1990a). Global change: translating plant ecophysiological responses to ecosystems. *Trends in Ecology & Evolution* **5**:308–310.

Woodward, F. I. (1990b). The impact of low temperatures in controlling the geographical distribution of plants. *Philosophical Transactions of the Royal Society London* B **326**:585–593.

Woodward, F. I., and A. D. Diament. (1991). Functional approaches to predicting the ecological effects of global change. *Functional Ecology* **5**:202–212.

Wright, I. J., and M. Westoby. (2001). Understanding seedling growth relationships through specific leaf area and leaf nitrogen concentration: generalisations across growth forms and growth irradiance. *Oecologia* **127**:21–29.

Wright, I. J., P. B. Reich, and M. Westoby. (2001). Strategy shifts in leaf physiology, structure and nutrient content between species of high- and low-rainfall and high- and low-nutrient habitats. *Functional Ecology* **15**:423–434.

Wright, I. J., P. B. Reich, M. Westoby *et al.* (2004). The worldwide leaf economics spectrum. *Nature* **428**:821–827.

Wright, S. J. (1967). Comments of the preliminary working papers of Eden and Waddington. *In* Moorhead, P. S., and M. M. Kaplan, editors. *Mathematical Challenges to the Neo-Darwinian Interpretation of Evolution*. Wistar Institute Symposium Monograph 5. Wistar Institute Press, Philadelphia, pp. 117–120.

Yee, T. W., and N. D. Mitchell. (1991). Generalized additive models in plant ecology. *Journal of Vegetation Science* **2**:587–602.

Zillio, T., and R. Condit. (2007). The impact of neutrality, niche differentiation and species input on diversity and abundance distributions. *Oikos* **116**:931–940.

Index

Printed in the United States
by Baker & Taylor Publisher Services